Protein Structure
New Approaches to Disease and Therapy

Max Perutz

Medical Research Council
Laboratory of Molecular Biology
Cambridge, England

 W. H. Freeman and Company
New York

The cover image was generated using UCSF MIDAS-plus
software. Modification of the created image was done by
B. de Vos and Dr. T. Hynes.

Library of Congress Cataloging-in-Publication Data

Perutz, Max F.

    Protein structure : new approaches to disease and therapy / M. F.
      Perutz
      p.   cm.
    Includes bibliographical references and index.
    ISBN 0-7167-7021-0 (hard)   ISBN 0-7167-7022-9 (soft)
    1. Proteins—Structure. 2. Pathology, Molecular. I. Title.
    [DNLM: 1. Crystallography—methods. 2. Proteins—physiology.
    3. Proteins—ultrastructure. QU 55 P471p]
    QP551.P398 1992
    612′.01575—dc20
    DNLM/DLC
    for Library of Congress   92-928
                                                      CIP

Printed in the United States of America

1 2 3 4 5 6 7 8 9 0 VB 9 9 8 7 6 5 4 3 2

How manifold are Thy works, Oh God,
Who can fathom their number?
Chorus in Haydn's Creation

To Sir Harold Himsworth, Secretary of the
British Medical Research Council from 1949 to
1968, whose foresight and courage led him to
support my colleagues' and my early work on
protein structures when there was only the
faintest hope of it ever benefitting medicine.

Contents

Preface

I began X-ray analysis of hemoglobin, the easiest protein to crystallize, as a graduate student in Cambridge, England, in 1937, because at the time the structure of proteins seemed to me the most fundamental unsolved problem in biochemistry and X-ray crystallography the only method capable of solving it. I was supported at first by my father and later by the Rockefeller Foundation, whose Director of Natural Sciences, Warren Weaver, originally coined the term *molecular biology* in his report to the President in 1939. After World War II, the Foundation felt that British sources should shoulder my project, but Cambridge University showed no interest. Fortunately I had the backing of Sir Lawrence Bragg, the prestigious pioneer of X-ray analysis and Cavendish Professor of Physics, who approached the Medical Research Council.[1] He warned the Council that there was only the remotest chance of success, but they decided to take the risk. Was it justified?

The first protein structures revealed wonderful new faces of nature, but they did not help to cure anyone. As far as practical benefits to medicine go, it always remained "jam tomorrow." When did it begin to be "jam today"? For me, the turning point came with Herman Waldmann's and Greg Winter's humanized rat anti–T-cell antibody that induced prolonged remissions in two terminally ill leukemia patients (Riechmann et al., 1988). Waldmann and Winter could never have engineered that antibody if others before them had not solved the structures of several immunoglobulins using X-ray analyis.

---

[1]The Medical Research Council is the British counterpart of the American National Institutes of Health.

I was overjoyed that protein crystallography had at last come of age, and I started to write a review about its medical applications. However, I soon found so much to write about that my review expanded into this book.

Many years ago, I began giving a course in X-ray crystallography of biologically important molecules for students of biochemistry and other biomedical subjects. In my first lecture I introduced lattice theory, trigonometric functions, and Fourier series very lucidly, I thought, but half the students failed to turn up for my second lecture. That sobering experience made me look around for other ways of explaining X-ray diffraction. The following year I replaced my forbidding lecture with a non-mathematical, largely pictorial introduction called "Diffraction Without Tears." This time my audience stayed, and I have now expanded that lecture into the first chapter of this book. I hope that it will make the basic principles of X-ray crystallography intelligible to non-physicists. For those familiar with physics, Appendix 1 gives a short introduction to the mathematical theory of X-ray analysis. I expect that most readers will already be familiar with the principles of protein and nucleic acid chemistry and structure, but novices will find these summarized in Appendix 2.

In 1953 I discovered that the phase problem of protein crystallography could be solved by the method of isomorphous replacement with heavy atoms. At the time I expected that the structures, not only of hemoglobin, but also of many other proteins, would soon be solved. This did not happen. Only three protein structures had been solved by 1965, and only eleven by 1970. The practical difficulties of crystallization, of preparing isomorphous heavy atom derivatives, and of recording the X-ray diffraction data were so great that determination of each new structure took many years. Besides, most professional crystallographers were reluctant to enter this risky new field. Today the situation is transformed. Since 1975 there has been an exponential rise in the annual number of protein structures solved; in 1990 alone over a hundred new ones came to light and by mid-1991 about 300 protein structures had been solved, many of them of practical interest to medicine, an interest that often became apparent only *after* they had been solved; sometimes years afterward (Fig. P.1).

The Human Genome Project has aroused great interest in medical circles, but locating the gene responsible for a disease is only the first step. For diagnosis of the true cause of the disease and an approach toward a rational treatment, it is essential to know the nature and

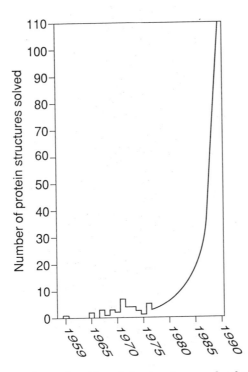

Figure P.1. Annual number of protein structures solved 1959–1990. (Courtesy of Dr. Authur Lesk)

function of the protein that is coded for by the gene. In order to design a drug, one also has to know its structure. For example, knowing that the gene for Huntington's Disease lies near the telomere of chromosome 4 has made it possible to identify carriers of the disease, but so far it has not brought diagnosis of its cause or treatment any nearer. On the other hand, the discovery that the gene for muscular dystrophy codes for the hitherto unrecognized protein dystrophin has already stimulated attempts at somatic gene therapy in children affected by the disease. As long as the structure of the HIV transcriptase, the enzyme that replicates the genome of the AIDS virus, was unknown, there was no rational way of improving on AZT, its only effective and merely moderately toxic inhibitor. Even the outline of this enzyme that is now emerging from the X-ray analysis of Tom Steitz and his colleagues at Yale University has already generated ideas for new antiviral drugs (see Chapter 10). From a medical point of view, therefore, protein chemistry and protein structure are essential complements of DNA sequencing.

I hope that this book will help research workers, teachers, and students to become familiar with, and put to practical use, the great body of new knowledge of molecular anatomy, physiology, and pathology that is now being created.

I wish to thank the following for reading either part or the whole of my manuscript and making helpful suggestions and criticisms: Professor P. W. Atkins, Dr. Cyrus Chothia, Dr. Fred Cohen, Dr. Mair Churchill, Dr. Tony Crowther, Dr. David Eisenberg, Dr. John Finch, Dr. Richard Henderson, Dr. T. Kouzarides, Dr. Terry Rabbitts, Dr. Daniella Rhodes, Dr. John Schwabe, Dr. Masashi Suzuki, and Dr. J. V. Watson. Dr. Crowther also calculated the Fourier transform in Figure A1.1 and drafted the accompanying text. I also wish to thank the many authors, editors, and publishers for their permissions to reproduce figures that appeared in books and journals. I particularly thank Dr. B. de Vos and Dr. A. Kossiakoff for supplying the then still unpublished picture of the human growth hormone with its receptors reproduced on the cover of this book; Dr. Arne Strand and Dr. Elizabeth Goldsmith for making the pictures of ovalbumin and plasminogen activator inhibitor shown in Figure 2.19 specially for me; Dr. Daniel Carter for sending me the picture of human serum albumin shown in Figure 10.1 in advance of publication; and Dr. Arthur Lesk for compiling the data and drawing the histograms of the number of protein structures solved over the years that enabled me to draw Figure P1. Several other authors kindly helped me to make my book topical by sending me manuscripts of their as yet unpublished papers. Finally, I wish to thank Mrs. Christine Strachan for helping me to bring this book into being.

1
Introduction

Diffraction Without Tears: A Pictorial Introduction to X-Ray Analysis of Crystal Structures

Relation Between Diffraction Pattern and Image

In 1934 J. D. Bernal and Dorothy Crowfoot (now Hodgkin) at the Crystallograhic Laboratory in Cambridge, England placed a crystal of pepsin in an X-ray beam to see if it gave a diffraction pattern. It was an unpromising experiment because it had already been proven that protein crystals give no diffraction pattern. This was only to be expected because the great German chemist Richard Willstätter and his pupils had shown that proteins are colloids of random structure, and that the enzymatic activity of J. H. Northrop's crystalline pepsin did not reside in the protein, which was but the inert carrier for its real, yet to be isolated, active principle (Waldschmidt-Leitz, 1933; Dyckerhoff and Tewes, 1933). Besides, even if the German chemists were wrong, and a diffraction pattern were obtained, it would clearly be impossible to deduce from it structures of molecules as large and complex as proteins.

Contrary to all reason, or perhaps because they had not read the literature, Bernal and Crowfoot discovered that pepsin crystals *did* give an X-ray diffraction pattern (Bernal and Crowfoot, 1934). It was made up of sharp reflections that extended to spacings of the order of interatomic distances, showing that pepsin was not a colloid of random coils, but an ordered three-dimensional structure in which most of its 5,000 atoms occupy definite places. Their observation opened the subject of protein crystallography.

The X-ray study of proteins is sometimes regarded as an abstruse subject comprehensible only to specialists, but the basic ideas underlying our work are so simple that some physicists find them boring (Holmes and Blow, 1965; Blundell and Johnson, 1976). Crystals of hemoglobin and other proteins contain much water and, like living tissues, they tend to lose their regular, ordered structure on drying. To preserve this order during X-ray analysis, crystals are mounted wet in small glass capillaries. A single crystal is then illuminated by a narrow beam of monochromatic X-rays (Fig. 1.1). If the crystal is rotated in certain ways, the spots can be made to appear in a series of concentric circles; within each circle they lie at the corners of a regular lattice that is related inversely to the arrangement of the molecules in the crystal (Fig. 1.2). The larger the molecules and the further they lie apart, the closer the spots are crowded together on the X-ray diffraction picture. Each spot has a characteristic intensity that is determined, in part, by the arrangement of atoms inside the molecules. The reason for the different intensities is best explained in the words of W. L. Bragg, who founded X-ray analysis in 1913—the year after Max von Laue had discovered that X-rays are diffracted by crystals. In his Nobel lecture he said:

> It is well known that the form of the lines ruled on a [diffraction] grating has an influence on the relative intensity of the spectra which it yields. Some spectra may be enhanced, or reduced, in intensity as compared with others. Indeed, gratings are sometimes ruled in such a way that most of the energy is thrown into those spectra which it is most desirable to examine. The form of the line on the grating does not influence the positions of the spectra, which depend on the number of lines to the centimetre, but the individual lines scatter more light in some direction than in others, and this enhances the spectra which lie in those directions.
>
> The structure of the groups of atoms which composes the unit of the crystal grating influences the strength of the various reflections in exactly the same way. The rays are diffracted by the electrons grouped around the center of each atom. In some directions the atoms conspire to give a strong scattered beam, in others their effects almost annul each other by interference. The exact arrangement of the atoms is to be deduced by comparing the strength of the reflections from different faces and in different orders. (Bragg, 1922)

Thus there should be a way of reversing the process of diffraction, of proceeding backwards from the diffraction pattern to an image of the arrangement of atoms in the crystal. The way to do this is best explained by recalling the German physicist Ernst Abbe's theory of microscopic vision. In an object illuminated by parallel light, interference between rays scattered by different parts of the

Figure 1.1. Crystal of hemoglobin (black dot) mounted in thin-walled quartz capillary between columns of 2 M ammonium sulfate solution. It is on a goniometer head that allows it to be tilted in two mutually perpendicular directions in order to make one of its axes parallel to the axis oscillation used to obtain the X-ray diffraction photograph shown in Figure 1.2.

object can be shown to give rise to diffraction. In some directions the scattered rays reinforce each other, while in others they are extinguished by interference. If the object is an optical grating, then the different orders of its diffraction pattern can be seen in the back focal plane of the objective after removing the eyepiece. Light emanating from the diffraction pattern is again subject to

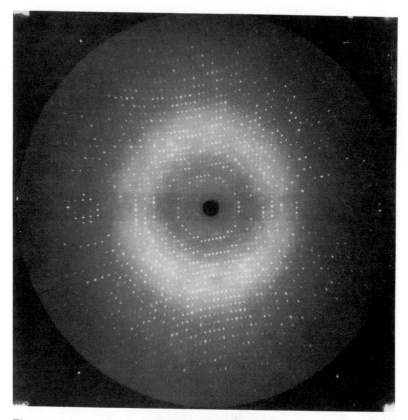

Figure 1.2. X-ray diffraction photograph of a hemoglobin crystal showing near the edge distinct reflections from lattice planes spaced as little as 1.75Å apart, which proves that equivalent atoms occupy closely similar positions in all the hemoglobin molecules throughout the crystal.

interference (Fig. 1.3). Provided the microscope is correctly focused, that interference reconstitutes a true magnified image of the grating. Hence there is a reciprocal relationship between an object, its diffraction pattern, and its magnified image. That relationship forms the basis of X-ray crystallography.

The Meaning of Phase

The intrinsic difficulty of that relationship becomes apparent if we consider what happens when the microscope is out of focus. This gives rise to a spurious image, because the different orders of

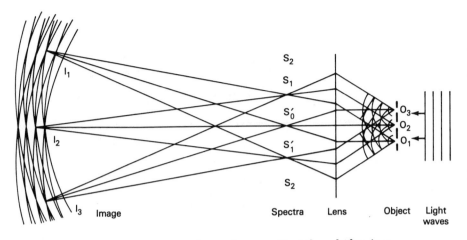

Figure 1.3. Generation of image of a grating, or of just three holes, in a microscope. S_0, S_1, S_2, etc. represent the zero, first, and second order diffraction spectra in the back focal plane of the objective.

the diffraction pattern are recombined with the wrong phases. Since X-rays cannot be focused by lenses, we cannot build an X-ray microscope, at least not yet. We can only record the intensities of the X-rays diffracted by a crystal, without being able to recombine them optically; it is as if we had a microscope without an eyepiece and did not know how to bring it to a focus. All we could do would be to take a photograph of the diffraction pattern. The image of the crystal structure has to be obtained by calculation, but since we cannot measure the phases directly, we have no direct way of reconstituting the diffraction pattern correctly and calculating the right image rather than a spurious one.

The meaning of phase can be illustrated by the diffraction of light shining through two closely spaced pinholes (Fig. 1.4). Let d be the distance between the holes, λ the wavelength, and α the angle between the diffracted ray and the normal to the line connecting the holes. In the straight forward direction the two waves are in phase and reinforce each other. When $\sin \alpha = \lambda/2d$ the rays scattered by the two holes are out of phase and annul each other. At $\sin \alpha = \lambda/d$, $2\lambda/d$, $n\lambda/d$ (where n is an integer giving the order of diffraction) they are in phase and reinforce each other. If the resultant wave is traced back to the midpoint between the two holes, there is a difference in phase between odd and even orders of diffraction. When n is odd, as in Figure 1.4C, the resultant wave has

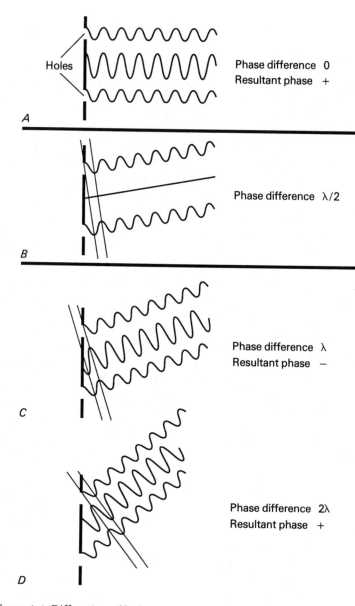

Figure 1.4. Diffraction of light waves scattered by two closely spaced pinholes. Interference extinguishes the light when the phase difference between the two waves is λ/2, 3λ/2, 5λ/2, etc. and reinforces it when the phase difference is an integral multiple n of λ. When n is odd, the resultant phase is negative; when n is even, it is positive. The oblique lines indicate wave fronts.

a minimum at that point, and we say that the phase is negative; when n is even, as in Figure 1.4D, it has a maximum there, and the phase is positive. Formation of the correct image depends on recombining the diffracted rays with their correct phases, but we cannot tell how to do this from just a photograph of the diffraction pattern.

Phase Determination by Isomorphous Replacement, and
Regeneration of the Image from the Diffraction Pattern
The difficulty can be overcome by the method of isomorphous replacement, which is best explained with the help of a simple optical device that tries to mimic an X-ray microscope. It was invented by the originator of X-ray analysis, W. L. Bragg, and it consists of a source of monochromatic light A, a pinhole B, two plane-convex lenses C and D, a mirror E, and a microscope (Fig. 1.5). The object is a mask placed between the lenses and perforated with holes that represent atoms. Its diffraction pattern is focused at F and viewed through the microscope. We take as our object hexamethylbenzene. Its diffraction pattern, shown in Figure 1.6A, can be thought of as due simply to the superposition of six sets of fringes, each of them arising by interference of light scattered by opposite pairs of holes. Suppose we were using X-rays rather than light. How then could we regenerate a picture of hexamethylbenzene from its diffraction pattern? For this purpose we have to know the phase of each of the bright areas. Since the molecule is symmetric about its center, the same symmetry must hold for each set of fringes; they must have either a peak or a trough at the center of the photograph, which means that the phases can be either plus or minus. The diffraction pattern is therefore made up of areas of positive and negative amplitudes separated by nodes where the amplitude passes through zero.

We now perform a thought experiment by introducing a thirteenth atom at the center of the benzene ring. The light scattered by it makes a positive contribution over the entire picture, thus making the positive areas brighter and the negative ones dimmer (Fig. 1.6B). We can now regenerate an optical image of hexamethylbenzene by cutting out a mask with holes of the same shape as the areas in the diffraction pattern, leaving the holes that correspond to positive areas open and covering the negative ones with a thin mica plate giving half a wavelength's retardation.

In practice we cannot observe the diffraction pattern of a single molecule, but only that of many in an ordered crystalline array, or disordered in a liquid or vapor, but we can at least approach the crystal problem in easy steps by starting with two molecules of

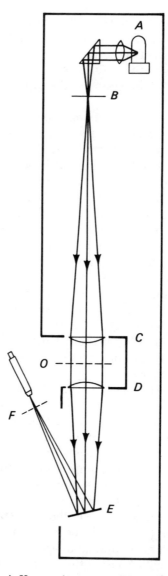

Figure 1.5. W. L. Bragg's X-ray microscope. For explanation see text.

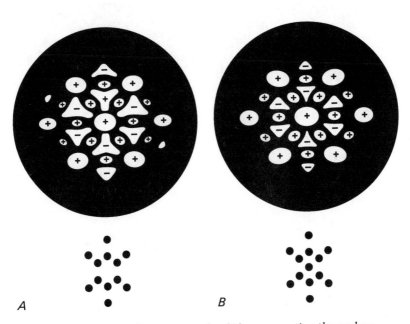

Figure 1.6. Diffraction from two masks: (A), representing the carbon atoms of hexamethylbenzene and (B), the same with an extra pinhole at the center of the benzene ring. Areas with positive phase have swollen in size, while areas with negative phase have shrunk. (Courtesy of Professor C. A. Taylor)

hexamethylbenzene placed side by side (Fig. 1.7). The bright spots now appear in the same positions as before, but they are intersected by a set of fringes perpendicular to the line separating the two molecules. The spacing between successive fringes is inversely proportional to that distance. Thus the diffraction pattern arises from the superposition of the interference fringes from a single molecule of hexamethylbenzene, intersected by those from two pinholes placed at the same distance apart as the two molecules. The result exemplifies a theorem according to which the diffraction pattern of any regularly repeating object arises form the convolution of the diffraction pattern of the single object with that of a set of point scatterers repeating at the same distance as that object. We shall see that this theorem leads to a simple interpretation of the X-ray diffraction pattern from double helices of DNA.

If four molecules of hexamethylbenzene are placed at the corners of a parallelogram, then the diffraction pattern of the single

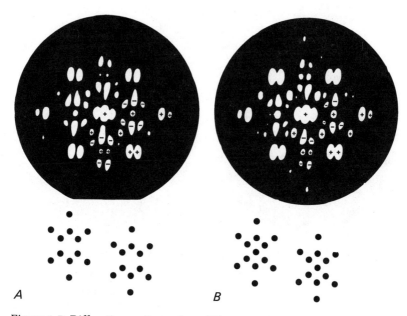

Figure 1.7. Diffraction patterns from (A), two molecules of hexamethylbenzene and (B), the same with an extra pinhole at the center of each ring. Note that the swollen positive areas and the shrunken negative ones overlap with those in Figure 1.6. (Courtesy of Professor C. A. Taylor)

molecule is intersected by two sets of interference fringes, each perpendicular to two of the four lines connecting the corners of the parallelogram (Fig. 1.8).

If the parallelograms are repeated at regular intervals in all directions, making a two-dimensional crystal, then the same fringes become sharper, and the diffraction pattern is reduced to bright spots at the intersections of the fringes. The spots are arranged in parallelograms whose sides are perpendicular to those of the real one; the lengths of their sides are inversely proportional to the perpendicular distance between the sides of those of the real lattice. In other words, the diffraction pattern of a regular two-dimensional array of molecules forms a lattice that is the reciprocal of the real lattice. The diffraction pattern of the single molecule is now no longer seen as a continuum of bright areas separated by nodes, but is still there all the same. It can be shown to be decomposed into the diffraction pattern from a regular grid of pinholes as in Figure 1.9, superimposed on the diffraction pattern from a single molecule of

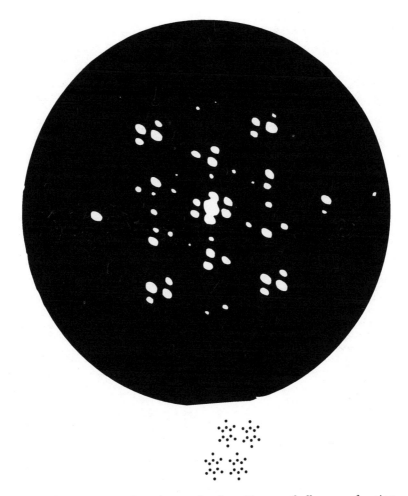

Figure 1.8. Diffraction from four molecules of hexamethylbenzene forming a parallelogram. The diffraction pattern of the single molecule is sampled at the crossing points of two sets of fringes as shown in Figure 1.9. (Courtesy of Professor C. A. Taylor) Figure A1.1 on page 254 gives a simple one-dimensional illustration of the fundamental principle that the diffraction pattern from a regular, periodic array of objects or apertures samples the diffraction pattern from the single object at discrete points.

hexamethylbenzene as in Figure 1.6A. *The diffraction pattern from the grid samples the amplitudes and phases of the pattern from the single molecule.* The phases can be determined as before by placing a thirteenth atom at the center of each molecule, which will

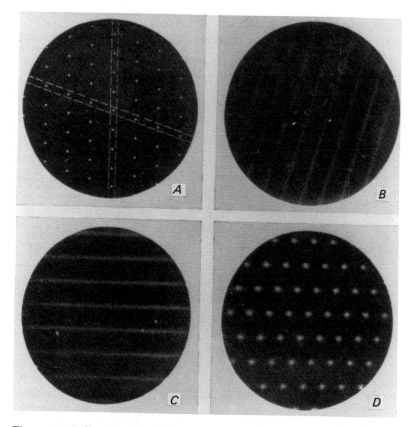

Figure 1.9. Diffraction from regular grid of pinholes. (A) The grid.
(B) Fringes diffracted by the one near-horizontal row of holes. (C) Fringes
diffracted by the one vertical row of holes. (D) Diffraction by the entire
grid. Light rays appear only where the two sets of fringes cross; they
form a lattice of points that is the inverse or reciprocal of the real lattice,
because the rows of lights are at right angles to the rows of holes and the
distances between neighboring lights are proportional to λ/d where λ is
the wavelength and d the perpendicular distance between rows of holes.
(Reproduced, by permission, from Bragg, 1975)

make spots with positive phase brighter and spots with negative
phase weaker.

An image of the structure can be generated by punching into a
mask holes of areas proportional to the brightness of each of the
spots, leaving holes with positive phase open and covering those
with negative phase with a half-wavelength plate. Each symmetry-
related pair of holes on either side of the center generates a set of

diffraction fringes. If the holes are numbered by the indices h and k along two mutually perpendicular axes x and y, then each set of fringes can be assigned indices h,k according to the pair of holes that generated it (Fig. 1.10). Note that the fringes from the closest pair of holes are furthest apart and vice versa. Hence the resolution of the final image improves with the orders of the fringes used to generate it.

Figure 1.11 illustrates why correct phases are essential for a correct image. Images are produced by superposition of fringes from a horizontal, a vertical, and an oblique pair of pinholes. Reversal of the phase of the fringes from the oblique pair radically changes the image.

Figure 1.12 shows how the image of diketopiperazine becomes progressively more blurred as the order of the X-ray reflections used to generate it is restricted. If all observed reflections extending to a spacing of 1.1Å are included, then every atom is pinpointed precisely. On the other hand, if all but an inner circle of reflections are excluded, then the image looks as if it had been painted with a very coarse brush. Looking back at Figure 1.10, the best image would have

Figure 1.10. Left: Diffraction pattern with intensities of diffracted rays being represented by circular holes of different areas. Orders of diffraction are numbered by the indices h along the x axis and k along the y axis. Right: Fringes produced by interference between pairs of rays from holes placed symmetrically about the center. The orders of the fringes correspond to the orders of the pairs of holes that produced them. (Reproduced from Beauclair, 1949)

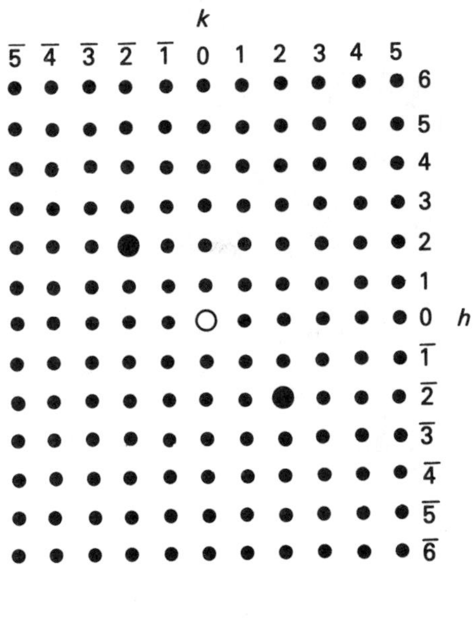

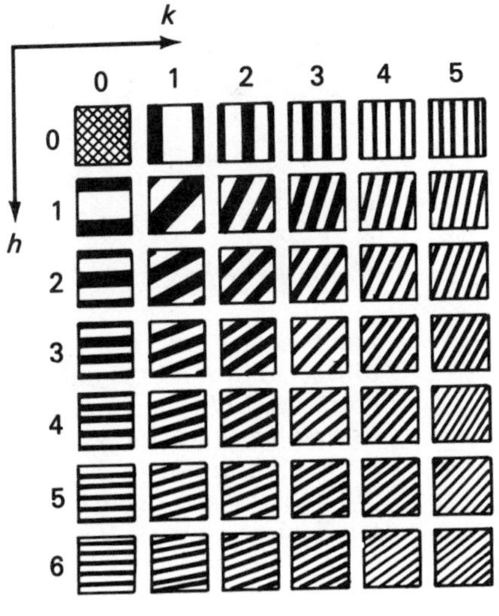

Wave III: $A_{11}\left\{1 + \cos\left[2\pi\left(\dfrac{x}{a} + \dfrac{y}{b}\right) + \alpha\right]\right\}$.

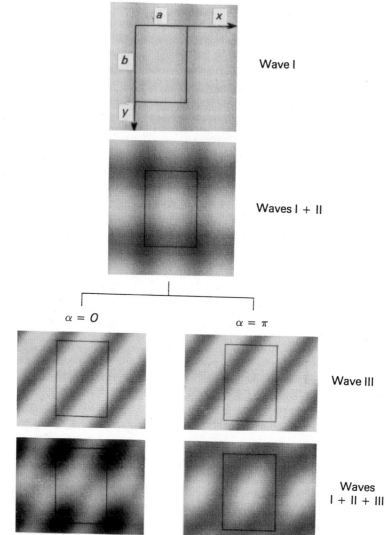

Figure 1.11. Optical superposition of diffraction fringes. Wave I is diffracted by a horizontal row of pinholes, wave II by a vertical one, and wave III by an oblique one. When the phase angle of wave III $\alpha = 0$, its first fringe has a crest at the top left hand corner of the image; when $\alpha = \pi$, it has a trough. Superposition of wave III on I + II produces different images, depending on α. For definition of phase angle see Fig. 1.15 on page 19. (Reproduced, by permission, from A. Guinier)

been generated by including fringes up to $h = k = \pm 6$; the worst by excluding all except those with $h = k = \pm 2$. (See also Fig. A1.1.) Such lack of resolution was one of our predicaments in the early days of protein crystallography.

The method of isomorphous replacement was first used in 1936 for the solution of what was then a complex structure, that of the dye phthalocyanin, by the Scottish crystallographer J. M. Robertson.

Figure 1.12. Series of electron density maps of diketopiperazine at diminishing resolution. 2Å is often the best resolution obtainable from protein crystals.

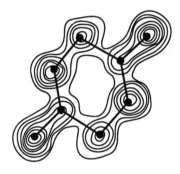

1.1Å resolution

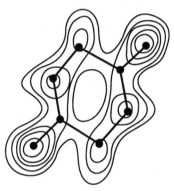

1.5Å resolution

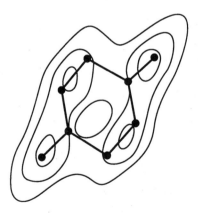

2Å resolution

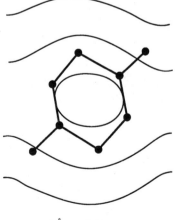

6Å resolution

Phthalocyanin had the virtues of being symmetric and of forming a series of isomorphous crystals with different metals chelated at its center of symmetry. Comparison of the X-ray diffraction from crystals of pure and of nickel phthalocyanin enabled Robertson to determine the phases and to calculate an electron density map of the molecule projected on a plane (Robertson, 1936). A few years later Robertson and Woodward determined the structure of the platinum derivative directly, without even having to use isomorphous replacement. They were able to this, because the scattering contributions from the 78 electrons of the platinum atom outweighed the scattering contributions from all the light atoms, many of which were attenuated by interference. Consequently all the fringes had a positive sign at the center of the platinum atom, which was also the center of symmetry of the light atoms. The electron density map calculated with all phases positive shows the carbon and nitrogen skeleton of the heterocyclic rings conjured up by pure physics, as though organic chemistry had never been invented (Fig. 1.13). In 1939 this success inspired J. D. Bernal to suggest that the same method might solve the structure of proteins, but at that time it seemed that the scattering contribution from a single heavy atom would make no measurable difference to the scattering contributions from the thousands of light atoms of a protein (Bernal, 1939).

In 1952 I happened to compare the absolute intensity of what seemed a very strong reflection from one of my hemoglobin crystals with the intensity of the incident beam. I found it to be much weaker than I had expected. The result made me realize that most of the scattering contributions by the light atoms are extinguished by interference, because these atoms are distributed more or less randomly over a large volume. It then occurred to me that the electrons of a heavy atom, being concentrated in a small sphere, would scatter in phase, and that their contribution should produce measurable intensity changes in the diffraction pattern of the protein. On the other hand, it was clear that these changes would provide the correct phases only if attachment of the heavy atom left the structure of the protein molecules and their arrangement in the crystal unaltered. When I first tried the method I was not at all sure that these stringent demands would be fulfilled, and as I developed my first X-ray photograph of mercury hemoglobin my mood alternated between sanguine hopes of immediate success and desperate forebodings of all possible causes of failure. I was jubilant when the diffraction spots appeared in exactly the same position as in the mercury-free protein, but with slightly altered intensities, just as I had hoped. Figure 1.14 shows

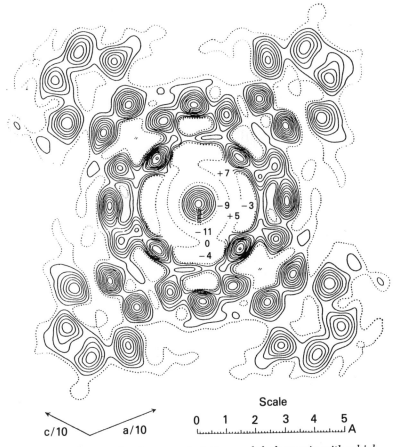

Figure 1.13. Electron density map of platinum phthalocyanin with a high peak representing the metal ion at the center. The metal is bound to four nitrogens, indicated by the four peaks that are closest to it; the remaining atoms are carbons. (Reproduced, by permission, from Robertson and Woodward, 1940)

the effect of substituting two atoms of mercury for two hydrogen atoms on a single row of X-ray reflections from a hemoglobin crystal.

Determination of Complex Phases: Double Isomorphous Replacement
So far we have demonstrated the use of isomorphous replacement only for the centrosymmetric molecule hexamethylbenzene. Being made up of enantiomorphous amino acid residues, protein molecules lack centrosymmetry, which means that the phases can have any value; they are complex. If we choose any arbitrary point in the

Intensities of 16 O L reflexions

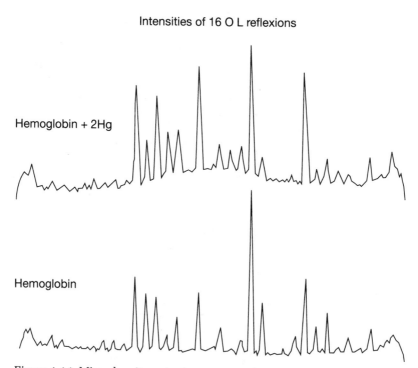

Figure 1.14. Microdensitometer trace across the same row of X-ray reflexions from two different crystals of hemoglobin, one without and the other with mercury atoms attached to the two reactive cysteines. Note the changes in intensity of equivalent reflections.

protein as an origin, diffracted rays may have either a crest or a trough or any intermediate phase at that point, and the changes in intensity produced by isomorphous replacement with a single heavy atom are no longer sufficient to determine the phases. The reason is as follows.

Suppose we represent the protein molecule by three atoms at the corners of an oblique triangle, and an X-ray scattered by that molecule as a sinusoidal wave (Fig. 1.15). We now introduce the heavy atom H_1, whose scattering contribution can be represented by another sinusoidal wave with a crest at H_1. Depending on their relative phases, the heavy atom wave may either add or subtract from the protein wave. From the difference in amplitude produced by the heavy atom, the distance of the crest of the protein wave from the heavy atom can be calculated.

Thus with the heavy atom serving as a common origin, the magnitude of the phase angle can be measured. However, this still

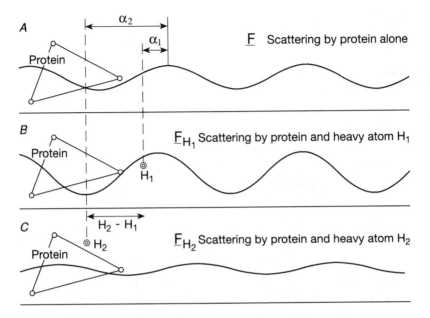

Figure 1.15. (A) Wave diffracted by triangle of atoms representing a protein. (B) and (C) Changes in amplitude and phase of diffracted wave caused by the heavy atoms H_1 and H_2. The phase is expressed as an angle α, taking the distance between successive wave crests as 360°. For example, in this figure the distance of the protein wave crest from H_1 is 7 mms. Successive wave crests are 37 mms apart. Hence the phase of the protein wave, measured from H_1 as the origin is $\alpha = \dfrac{7}{37} \times 360 = \pm\, 68$°.

leaves an ambiguity of sign, because the experiment does not tell us whether the phase angle is to be measured from the heavy atom in the forward or the backward direction. If n is the number of diffracted spots, an ambiguity of sign in each set of fringes would lead to 2^n alternative images of the structure. The Dutch crystallographer J. M. Bijvoet pointed out in another context that the ambiguity could be resolved by examining the diffraction pattern from a second isomorphous heavy-atom compound (Bokhoven et al.,1949).

Figure 1.15C shows that the heavy atom H_2, which is attached to the protein in a position different from that of H_1, diminishes the amplitude of the wave scattered by the protein. The degree of attenuation allows us to measure the distance of the wave crest from H_2. It can be seen that the wave crest must be in front of H_1; otherwise its distance from H_1 could not be reconciled with its distance from H_2. Thus the ambiguity of sign can now be resolved, provided that the vector H_1–H_2 is also known. The determination of that vector is one of the vital and often difficult steps in the X-ray analysis. As long as the number of heavy atoms associated with each protein molecule is not too large, it can be determined from the intensity differences by the difference Patterson method described in Appendix 1.

Determination of Complex Phases: Anomalous Scattering
Bijvoet demonstrated another way of solving the sign ambiguity that requires only a single heavy atom. If the energy of the incident X-ray photon is slightly larger than that needed to lift one of the heavy atom's electrons from the S shell into either the K or the L shell, then that electron reverts to the S shell with the emission of an X-ray photon of slightly lower energy than that of the incident X-rays. This is known as anomalous scattering, to distinguish it from normal scattering with the same photon energy as that of the incident X-rays. The anomalously scattered wave has two components; one of amplitude f′ that is in phase with the normally scattered wave, and another of amplitude f″ with a phase of a quarter wavelength ahead of the normally scattered wave.

Depending on the sign of the phase angle, that second component either increases or decreases the intensity of a diffracted ray. Since the phase angles of reflections from opposite sides of the same crystal face have opposite signs, anomalous scattering increases the intensities of one and decreases that of the other (Fig. 1.16). The signs of these intensity changes provide the signs of the phase angles.

Alternatively, the X-ray diffraction pattern can be recorded at different frequencies of the X-rays, in the presence or absence of anomalous scattering. Yang, Hendrickson, and others used this

Figure 1.16. (opposite) Determination of the sign of the phase angle by anomalous scattering. (A) F_p *wave scattered by protein alone, with amplitude* | F_p | *and phase* α. $f_H + f_H''$: *wave scattered normally by heavy atom.* $F_p + f_H + f_H''$: *wave scattered by protein plus heavy atom. The change in amplitude produced by the heavy atom gives the magnitude of the phase angle* α, *but does not tell if it is to be measured backward from the heavy atom, as shown, or forward. (B) and (C) Observed diffraction* F_o *from opposite sides of the same set of lattice planes, showing the anomalously scattered component* f''_H *of the heavy atom wave. Its phase is always 90° ahead of the normally scattered heavy atom wave. As a result, the anomalous component adds to* $F_p + f_H + f_H''$ *in one direction,* $F_o(-h)$ *(C), and subtracts from it in the opposite direction,* $F_o(h)$ *(B). This is consistent only with* α *being negative.* F_o *is the observed amplitude.*

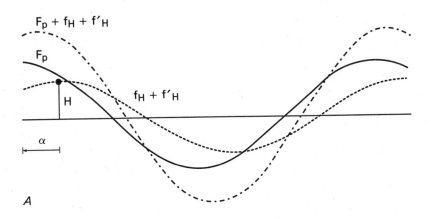

$F_p + f_H + f'_H$

F_p

$f_H + f'_H$

H

α

A

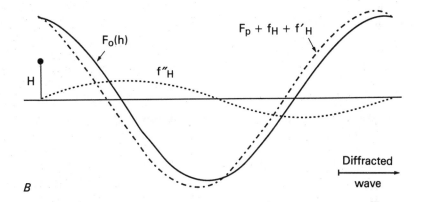

$F_0(h)$

$F_p + f_H + f'_H$

f''_H

H

Diffracted
wave

B

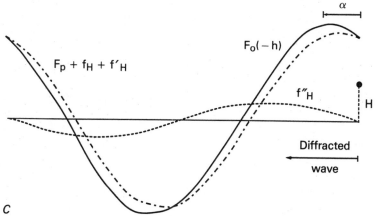

α

$F_0(-h)$

$F_p + f_H + f'_H$

f''_H

H

Diffracted
wave

C

method to determine the crystal structure of ribonuclease H of *E. coli*. They replaced its methionines with selenomethionines and recorded the X-ray diffraction pattern with a tunable X-ray source at three different frequencies. This allowed them to excite anomalous scattering by the selenium atoms and to measure F_H, f', and f'' for most reflections. In this way they avoided isomorphous replacement and determined the phases from just one crystal.

David Harker introduced a graphic method of calculating the phase angles which illustrates the heavy atom method very clearly. He used an Argand diagram in which the observed amplitude of the ray diffracted by the protein alone is represented by a vector of length $|F|$ and its phase by its angle α with the horizontal axis. If the heavy atom lies at the origin, the calculated amplitude of its scattering contributions is represented by the horizontal vector of length $|f_c + f_c'|$ plus a vector 90° clockwise to the horizontal of length $|f_c''|$. A circle of radius F is drawn around the origin. Another circle of radius F_H, the observed amplitude of the X-ray diffracted by the heavy atom derivative, is drawn around the end of the vectors $f_c + f' + f''$. Figure 1.17A shows how in the absence of anomalous scattering the angle between the horizontal and a vector drawn to the intersection between the two circles gives the phase angle α with an ambiguity of sign. Figures 1.17B and 1.17C show how that ambiguity can be resolved with the help of the anomalous scattering by the heavy atom. Finally, Figure 1.18 shows how the ambiguity can also be resolved with the help of two or more heavy atom derivatives when all the circles of radius F_H intersect at a single point. The angle between the horizontal and the vector to that point gives the magnitude and sign of the phase angle.

Diffraction in Three Dimensions: Bragg's Law and the Reciprocal Lattice

So far all the examples have been planar, as if we lived in the imagined two-dimensional world of Edwin Abbott's *Flatland*. In a three-dimensional crystal the interpretation of the diffraction pattern as a superposition of fringes becomes too cumbersome. In 1913, shortly after Max von Laue's discovery of X-ray diffraction, W. L. Bragg found a simpler way. In a crystal lattice, regularly spaced parallel planes can be drawn through equivalent points. If the planes are populated by atoms or molecules, they reflect X-rays like mirrors. Since X-rays penetrate crystals, reflections from successive lattice planes are extinguished by interference, unless the reflected rays are in phase and reinforce each other. If λ

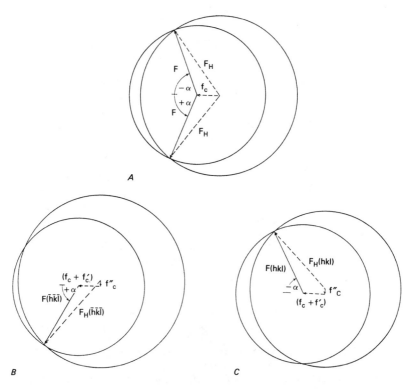

Figure 1.17. Representation of amplitudes of diffracted waves by an Argand diagram. Each wave is represented by a vector. The length of the vector represents the amplitude of the wave and its angle with the horizontal α the phase of the wave. (A) If the heavy atom is taken as the origin, the phase angle of its scattered wave is zero; hence the calculated heavy atom vector f_c is horizontal. F is the protein vector (called its structure factor), F_H the vector of the protein plus the heavy atom, which is the sum of the vectors $F + f_c$. F, F_H, and f_c form a triangle which provides the magnitude but not the sign of α. (B) and (C) Diffraction from opposite sides of the same crystal face in the presence of anomalous scattering by the metal atom. The sign of α is reversed, but f''_c, being 90° ahead of $f_c + f'_c$, enhances F_H when α is positive and weakens F_H when α is negative. In this way anomalous scattering by the heavy atom resolves the ambiguity of sign.

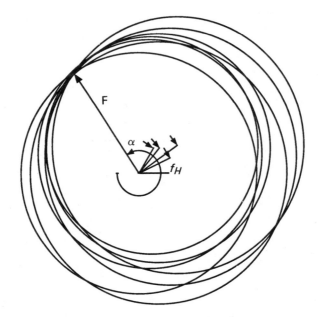

Figure 1.18. Determination of phase angle of protein structure amplitude F by multiple isomorphous replacement with heavy atoms attached to different sites of the protein. The heavy atom contributions are represented by the short vectors at the center. Circles with amplitude $F + f_H$ for the different derivatives are drawn around the ends of the short vectors. The angle between the horizontal and a vector drawn to the intersection of the circles gives α.

is the wavelength of the X-rays, d the perpendicular distance between successive planes, θ the angle between the planes and the incident or reflected X-ray, and n an integer, then reinforcement occurs only when $n\lambda = 2d \sin \theta$ (Fig. 1.19). This is known as Bragg's law.

With monochromatic X-rays and a stationary crystal, Bragg's law is met for only a few sets of planes, but rotation of the crystal makes reflections from different sets of planes flash up one by one. The X-ray diffraction picture of the hemoglobin crystal shown in Figure 1.2 was obtained by oscillating the crystal backwards and forwards by 1.25°.

Just as the diffraction pattern from regularly spaced parallelograms in a plane can be represented by points at the corners of reciprocal parallelograms, so the diffraction pattern from a crystal can be represented by points at the corners of regularly repeating parallelepipeds, which are the reciprocal of the parallelepipeds that form

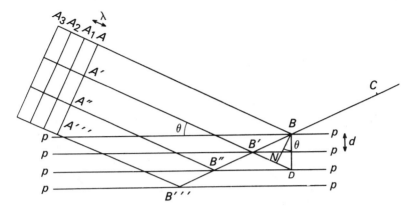

Figure 1.19. Diffraction of X-rays from a set of parallel lattice planes p, spaced at a distance d. A, A_1, A_2, and A_3 represent successive crests of the incident wave, BN the incident wave front at the point of reflection, θ the angle of incidence and reflection, and ND the phase difference between waves reflected at B and B', which is equal to 2d sin θ.

the unit of pattern in the crystal. We call the real parallelepiped the *unit cell* (Fig. 1.20). The axes of the reciprocal unit cell are drawn perpendicular to the three principal planes of the real unit cell, with length inversely proportional to the perpendicular distances between those planes (Fig. 1.21).

Figure 1.20. Left: Unit cell of hexamethylbenzene. Right: Crystal of hexamethylbenzene, showing the indices h, k, and l given to its three faces. (Reproduced, by permission from Bunn, 1945)

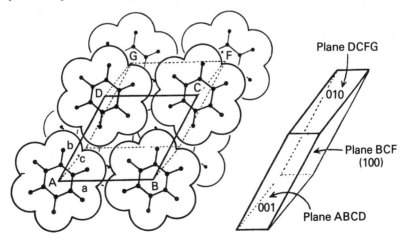

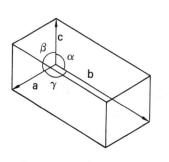
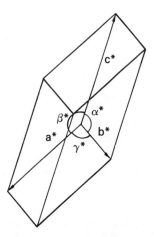

Figure 1.21. Relationship between real unit cell (left) and reciprocal unit cell (right). a* is perpendicular to the bc-plane, b* to the ac-plane, c* to the ab-plane. The lengths of a*, b*, and c* are inversely proportional to the distance between the respective planes in the real unit cell.

The diffraction pattern can be visualized as a sphere filled with layer upon layer of diffraction spots at the corners of the reciprocal lattice. The maximum radius of the sphere is determined by the wavelength of the X-rays used and by the Bragg equation: since $\sin \theta$ cannot exceed unity, the minimum spacing between two planes in the real lattice d_{min} must be $> \lambda/2$. With radiation from a copper anode, such as is widely used for protein crystals, $\lambda = 1.54\text{Å}$, but the X-ray patterns from protein crystals rarely extend to spacings of 0.77Å, because thermal motion and intrinsic disorder blur their structures.

The reciprocal lattice samples the diffraction pattern of single molecules or small groups of them. Figure 1.22 shows the diffraction pattern of naphthalene plotted on a stack of transparent sheets; each X-ray reflection is represented by a dark spot of an area proportional to its intensity. A view through the stack perpendicular to the plane of the naphthalene molecules in the crystal shows how the diffraction pattern from the single molecule enhances the intensity of some of the diffracted rays: it can be seen clearly as a hexagon of dark spots, similar to the hexagonal diffraction pattern from a single molecule of hexamethylbenzene. The lines from the center of the hexagon to the dark spots lie perpendicular to the sides of the naphthalene hexagons, and their distance from the center is inversely related to the distance between two sides of the hexagon.

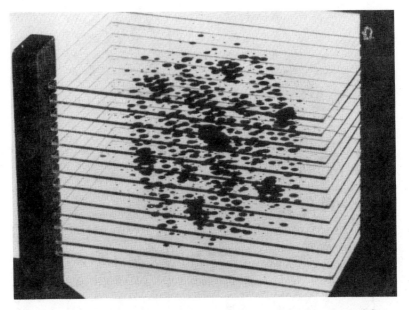

Figure 1.22. Reciprocal lattice of naphthalene crystal. The areas of the circles at each lattice point h,k,l are proportional to the amplitudes | F | of the X-ray reflection from the h,k,l plane. Note the dark areas at the corners of a hexagon. They lie at the ends of vectors drawn normal to the sides of the benzene rings. (Reproduced, by permission, from Bragg, 1975)

This is yet another example of the sampling of the diffraction from a single molecule by the diffraction pattern generated by a regular three-dimensional array of the same molecules.

An image of the crystal structure in three dimensions can be reconstructed from its diffraction pattern as before. Each pair of symmetrically related X-ray reflections is used to generate a set of fringes with the correct amplitude and a phase related to an arbitrary origin, say a heavy atom. Figure 1.23 shows such sets of three-dimensional fringes. All these fringes are then superimposed by calculation. This means that the unit cell is divided into perhaps $50^3 = 125,000$ compartments and the values of each of tens of thousands of sets of fringes are calculated and added for each compartment. The results can be plotted as a series of contour maps, like microtome sections through a tissue, only on a thousand-times-smaller scale. With small molecules like penicillin they delineate individual atoms (Fig. 1.24), but with proteins such high resolution is rarely attainable. Figure 1.25 shows an electron density

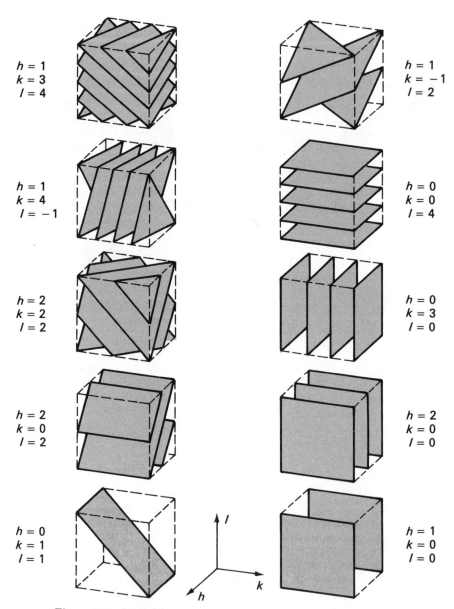

$h = 1$
$k = 3$
$l = 4$

$h = 1$
$k = -1$
$l = 2$

$h = 1$
$k = 4$
$l = -1$

$h = 0$
$k = 0$
$l = 4$

$h = 2$
$k = 2$
$l = 2$

$h = 0$
$k = 3$
$l = 0$

$h = 2$
$k = 0$
$l = 2$

$h = 2$
$k = 0$
$l = 0$

$h = 0$
$k = 1$
$l = 1$

$h = 1$
$k = 0$
$l = 0$

Figure 1.23. Sets of fringes corresponding to the terms of a Fourier series used to build up the three-dimensional image of a structure. The indices h,k,l are the three-dimensional equivalents of the planar indices of Figure 1.10. (Reproduced, by permission, from Beauclair, 1949)

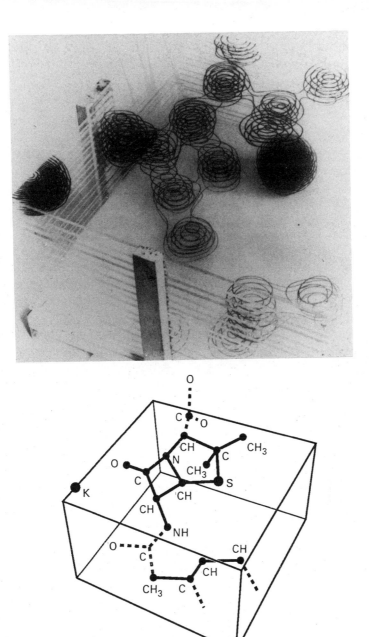

Figure 1.24. Three-dimensional electron density map of benzylpenicillin. From this map was deduced its chemical formula, which had resisted chemical methods of structure analysis. (Courtesy of Professor Dorothy Hodgkin)

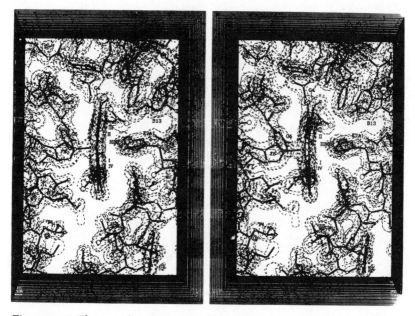

Figure 1.25. Electron density maps of human deoxyhemoglobin showing the α-heme in profile at 1.74Å resolution. (Reproduced, by permission, from Fermi et al., 1984)

map of human deoxyhemoglobin at 1.74Å resolution, the best that could be obtained in 1980. The heme is well resolved, and atoms can be placed with a root mean square error of about 0.2Å, compared to 0.01Å or better in crystals of small organic compounds.

In the X-ray diffraction patterns of protein crystals the number of spots runs into tens or hundreds of thousands. In order to determine the phase of each spot accurately, its intensity must be measured several times over: in the diffraction pattern from a crystal of the pure protein and in the patterns from crystals of several compounds of the protein, each with heavy atoms attached to different positions in the molecule. Then the results have to be corrected by various geometric factors before they are finally used to build up an image through the superposition of all the fringes. In the final calculation hundreds of millions of numbers may have to be added or subtracted. Such a task would have been quite impossible before the advent of high-speed computers, and crystallographers have been fortunate that the development of computers has kept pace with the expanding needs of their X-ray analyses, from myoglobin with about 1,200 atoms other than hydrogen in 1960, to the photochemical reaction center with about 12,000 atoms other than hydrogen in 1985. At the same time photographic methods of recording intensities have been supplemented by computer-controlled physical devices that measure the intensities automatically and feed them directly into memory stores. The heavy atom method has been supplemented by others. One of them exploits restrictions placed upon the phases by symmetry (for instance, in viruses that have multiples of 60 identical protein molecules distributed symmetrically over their surface). Another exploits the sometimes large water-filled spaces between the protein molecules that must remain featureless in the final map.

I have likened the diffraction pattern from a protein crystal to a hieroglyphic without a key, but my analogy limps because once the key had been found, the Rosetta stone was deciphered word for word. On the other hand, X-ray crystallographers toiling away at their structure year after tedious year receive no hint of any of its features until the final dramatic moment, when the complete electron density map emerges all at once and reveals a face of nature that no one has beheld before. After the first elation, the crystallographer faces the nagging question: Is the map right?

If the amino acid sequence of the protein has been determined by chemical methods, the question may answer itself. In 1967 when, after six years of hard work, my colleagues and I had raised the resolution of the electron density map of horse methemoglobin from its initial low value of 5.5Å to 2.8Å, the aromatic side chains stood out on the map like signposts in the wilderness. Some years earlier, D. B. Smith in Canada had determined the amino acid sequence of horse hemoglobin, but none of this information had entered into our X-ray analysis. The agreement between the positions of the aromatic residues predicted by the chemical sequences and their positions on the physical map dispelled any doubts about its correctness. It really made us believe in science!

Eisenberg and his colleagues have developed rigorous criteria for assessing whether the known chemical sequence of a protein has been fitted to the electron density map correctly. They assign positions of residues in accurately known protein structures to a series of classes, depending on each position's environment: (i) the total area of the sidechain that is buried by other residues; (ii) the fraction of sidechain buried by polar atoms or water; (iii) whether the residue at that position is part of an α-helix, a β-strand, a β-turn or an irregular loop. Having thus derived a table of environmental preferences for each of the twenty amino acids in known structures, they give each residue in the newly determined structure a score, depending on the consistency of its environment there with those established preferences. This method gave high scores to correct structures and low scores to structures that later proved wrong. Even local errors in interpretation of the electron density map seemed to show up as low scores (Bowie et al., 1991; Lüthy et al., 1992).

Diffraction from Helical Chain Molecules

One afternoon in 1952 I posed to two of my physicist colleagues, Bill Cochran and Francis Crick, the problem of interpreting the diffraction pattern of polybenzyl-L-glutamate which Pauling had recognized the previous year as coming from an α-helix. Each of them worked out the solution that same night and arrived in the laboratory next morning with the same answer, though they had worked it out in different ways. Cochran explained that he had convoluted the diffraction pattern of a helical wire with that from a series of parallel planes spaced 1.5Å apart perpendicular to the axis of the helix. His argument went along the following lines.

A helical wire gives rise to a pattern of diffracted spectra that lie on a cross. The axes of that cross are perpendicular to the zigzags of the helix as seen in projection on a plane parallel to the helical axis (Fig. 1.26A). Within that cross the spectra lie on layer lines spaced at a distance P^* that is inversely related to the pitch of the helix P. On each layer line the diffracted intensity varies according to a mathematical function, the Bessel function, named after the mathematician who first described it (Fig. 1.27). The intensity along the central layer lines varies according to the zero order Bessel function; that along the first layer line varies according to the first order, and so on.

Suppose that the helix is made up of single atoms repeating at an axial distance of p = 1.5Å in the vertical direction. This then is equivalent to the density of matter along the helix being zero everywhere except where it is intersected by a series of parallel planes spaced 1.5Å apart perpendicular to the helix axis. Such planes would produce reflections at an angle of $\sin \theta = {}^{n\lambda}/3.0 = np^*$ along the vertical axis of the photograph. Figure 1.26(B) shows the two first-order reflections at $+p^*$ and $-p^*$. Convolution of the two diffraction patterns means that the diffraction pattern from the helical wire of Figure 1.26A is repeated at every point $+p^*$ and $-p^*$ of the diffraction pattern from the set of parallel

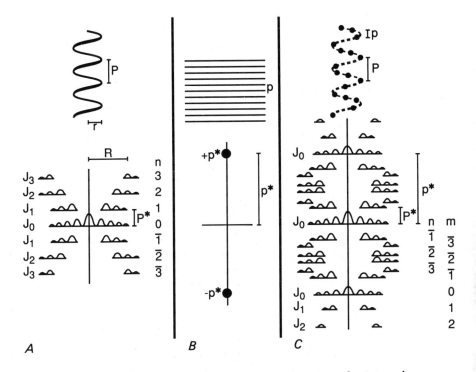

Figure 1.26. Diffraction from a single atom repeating at regular intervals
along a helix of pitch P. (A) Diffraction from a helical wire. (B) Diffraction
from a set of parallel planes perpendicular to the helix axis spaced pÅ apart.
(C) Convolution of diffraction patterns (A) and (B). J_0 to J_3 indicate the
orders of Bessel functions on the layer lines n=-3 to n=3, centered on the
equator; and those on the layer lines m=-3 to m=3, centered on +p* and −p*.

planes. In practice only the first orders of the latter are resolved, so
that the helical pattern is repeated just once on each side of the
center (Fig. 1.26C). This theory was crucial for Watson and Crick's
solution of the structure of DNA the following year.

    In this chapter I have tried to present a nonmathematical account
of the physical principles of X-ray analysis, as W. L. Bragg taught
them to me over the years that we worked together. Although his
first degree was in mathematics, his approach to science was
artistic, and he had a lucid, visual grasp of physical phenomena. I
hope that the chapter will help readers trained mainly in the
biological and medical sciences to understand the basic ideas of
X-ray crystallography. Its practical applications do indeed require

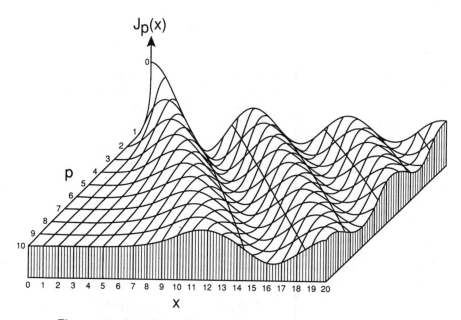

Figure 1.27. Set of Bessel functions of order 0, 1, 2, etc., corresponding to the distribution of diffraction amplitude on layer lines 0, 1 and –1, 2 and –2, etc. of diffraction from the helical wire in Figure 1.26.

formidable computations, but having performed them myself by hand before there were computers, I learned that they are largely repetitive. Appendix 1 introduces the mathematical concepts involved.

The determination of the structures of proteins and nucleic acids by X-ray analysis has received a tremendous boost by gene technology. Once their genes have been isolated and cloned, proteins previously extractable from biological tissues only in trace quantities of doubtful purity can now be made by the hundred milligrams in genetically transformed microorganisms or in mammalian cells. They can then be efficiently purified and crystallized. As a result, the number of protein structures solved since January 1988 probably exceeds that of the previous thirty years. Yet even today not all proteins can be crystallized and not all crystalline proteins yield detailed X-ray diffraction patterns.

Structures of Proteins in Solution

X-ray crystallography has been complemented by proton nuclear magnetic resonance (NMR), which has allowed structures of proteins of up to about 150 amino acid residues to be solved in solution. Like the X-ray diffraction pattern, the complex pattern of thousands of proton resonances from a macromolecule resembles a hieroglyphic without a key. That key lies in the assignment of each resonance to a specific proton in each individual amino acid residue (Wüthrich, 1986, 1989). Once that crucial problem of resonance assignment has been solved, the spectra can tell which pairs of protons are separated by distances of less than 5Å, A matrix of these short distances is then drawn up. This tells immediately if a segment of chain is α-helical or extended, since in an α-helix main chain protons four residues apart are separated by only about 4Å, while in an extended chain they would lie about 14Å apart. Similarly, a pleated β-sheet can be recognized by the systematic pattern of short distances between main chain protons in adjacent segments of polypeptide chain.

Next, the matrix is searched for resonances that tell which of those segments of α-helix or pleated sheet make contact with each other. Such qualitative information is helpful, but insufficient for an accurate structure. Instead, sophisticated algorithms based upon theorems of distance geometry are fed into powerful computers to calculate the exact fold of the main polypeptide chain and the conformations of the side chains consistent with the matrix of *all* the observed short distances between protons. However, that matrix contains a range of allowed distances for each pair of protons, which raises the question of how the structure changes as these distance are varied between, say, 1.4 and 5Å. This question is addressed by repeating the same calculation several times, each time with a different set of randomly chosen distances between neighboring protons within the allowed range. The result is an ensemble of related structures, each consistent with the observations. The differences between them are then expressed as the root mean square distance between the same atom in the different structures. That distance diminishes the greater the number of resolved resonances in the initial matrix, just as the accuracy of an X-ray structure improves with the number of measured reflections.

Figure 1.28 shows a typical result. The polypeptide hormone glucagon is represented by an ensemble of similar structures generated by repeating the same calculation with slightly different proton-proton distances randomly generated between an upper and a lower limit (Wüthrich, 1989). This structure of glucagon in solution was found to be different from that in the crystal. For a small peptide this was not unexpected, because the energy of the

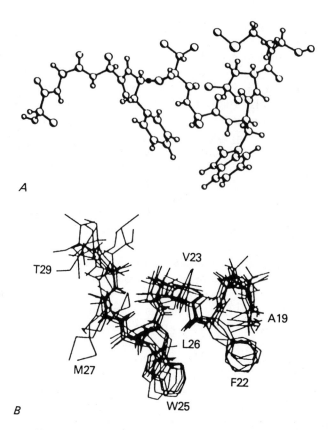

Figure 1.28. NMR structure of part of the peptide hormone glucagon at the lipid-water interface on the surface of dodecylphosphocholin micelles. (A) One of the structures calculated for the segment Ala19-Glu-Asp-Phe-Val-Glu-Trp-Leu-Met 27. (B) Six structures calculated from the same set of measurements are superimposed. (Reproduced, by permission, from Wüthrich, 1989)

noncovalent bonds formed between neighboring molecules in the crystal lattice is likely to exceed the energy of those formed between residues within the peptide. On the other hand, all structures of larger proteins in solution determined by NMR have proven to be closely similar to those found by X-ray crystallography. For example, a detailed comparison of the structures of the α-amylase inhibitor tendamistat determined independently by the two methods yielded root mean square deviations of atomic positions between the NMR and the X-ray structures of 1.05Å for the backbone atoms N,

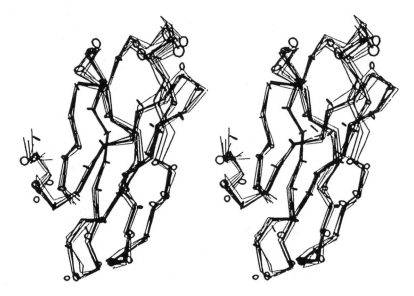

Figure 1.29. Outline in stereo of the α carbon skeleton of the α-amylase inhibitor tendamistat, showing residues 5-73. The α-carbon positions derived by X-rays are shown as small circles, while the lines represent the nine structures derived by NMR. The radii of the circles are proportional to the mean displacements of the atoms from their equilibrium positions, due to their thermal motion. (Reproduced, by permission, from Billeter et al., 1989)

Cα, and C'; 1.25Å for the entire backbone and the interior side chains; and 1.84Å for all nonhydrogen atoms (Fig. 1.29). Three external side chains had different conformations in solution and in the crystal, probably because they made contact with neighboring molecules in the crystal, and one tyrosine side chain that was not visible in the electron density map was resolved in solution. In general, there was a close correlation between the mobilities of different side chains found by the two methods. On the other hand, NMR has revealed some mobilities of which X-rays had given no hint. For example, in the bovine pancreatic trypsin inhibitor a buried phenylalanine sidechain was found to flip over several hundred times a second at room temperature, even though the phenyl ring is packed tightly against its neighbors. Such flipping can occur only if the entire protein molecule is in rapid and extensive thermal motion. NMR can also tell which NH protons are linked to oxygens by tight hydrogen bonds and which hydrogens are freely exchangeable.

Atomic nuclei with odd numbers of protons and neutrons have unpaired spins that give them large net magnetic moments, while those with even numbers have paired spins that almost cancel. Thus ^1H, ^{13}C, ^{15}N, ^{17}O, and ^{31}P all have large magnetic moments, while ^2H, ^{12}C, ^{14}N, ^{16}O, and ^{32}P have smaller moments. In proteins found in nature, therefore, only protons and not naturally abundant isotopes of other atoms have large magnetic moments. On the other hand, proteins synthesized in genetically engineered microorganisms can be enriched or even completely substituted with isotopically labelled amino acids. This possibility is now being exploited for NMR. Kay and others have prepared the 153 amino acid hormone human interleukin-1β with residues labelled with both ^{13}C and ^{15}N. As a result, each proton is shifted in frequency according to the precise magnetic moment of the ^{13}C or ^{15}N to which it is attached, and these frequencies in turn vary with those atoms' surroundings. The resulting shifts greatly facilitate the assignment of resonances to particular protons and also improve their resolution (Kay et al., 1990), allowing larger protein structures to be solved by NMR. (See Chapter 9.)

There has been one disturbing report of the structure of an enzyme, metallothionine, being different in solution and in the crystal (Vasák et al., 1987), but this was later found to be due to an erroneous interpretation of the X-ray data. When that was corrected, the two structures agreed.

To the newcomer the sheer complexity of protein structures looks as forbidding as the mathematics needed to solve them, but again the underlying principles are simple. Since most readers may be already familiar with them, I have relegated them to Appendix 2, together with a brief introduction to nucleic acids.

FURTHER READING

Diffraction Without Tears: A Pictorial Introduction to X-ray Analysis of Crystal Structures
Blundell, T.B., and L. H. Johnson 1976. *Protein Crystallography*. Academic Press, New York. An excellent introduction.
Bunn, C. W. 1945. *Chemical Crystallography*. Clarendon Press, Oxford, England. An old book, but still the best introduction for non-physicists.
Carter, C. W. 1990. Protein and nucleic acid crystallization. *Methods* 1:1–127. Academic Press, San Diego, CA. A useful collection of papers.
Dunitz, J. D. 1979. *X-ray Analysis and the Structure of Organic Molecules*. Cornell University Press, Ithaca, N.Y. An excellent introduction for chemists.

Glusker, J. P. 1981. *Structural Crystallography in Chemistry and Biology.* Hutchinson Ross Publishing Co., Stroudsburg, PA. A collection of historic papers.

Green, D. W., V. M. Ingram, and M. F. Perutz 1954. The structure of Haemoglobin IV: sign determination by the isomorphous replacement method. *Proc. Roy. Soc.* **A225**:287-307.

Holmes, K. C., and D. M. Blow 1965. The use of X ray diffraction in the study of protein and nucleic acid structure. *Methods of Biochem. Anal.* **13**:113–239. A concise and lucid account of the subject.

Kendrew, J. C., G. Bodo, H. M. Dintzis, H. Parrish, H. Wyckoff, and D. C. Phillips 1958. A three-dimensional model of the myoglobin molecule obtained by X-ray analysis. *Nature* **181**:662–666.

Kendrew, J. C., R. E. Dickerson, B. E. Strandberg, R. J. Hart, D. R. Davies, D. C. Phillips, and V. C. Shore 1960. Structure of myoglobin: a three-dimensional Fourier synthesis at 2Å resolution. *Nature* **185**:422-427.

Perutz, M. F., M. G. Rossmann, A. F. Cullis, G. Muirhead, G. Will, and A. T. North 1960. Structure of haemoglobin: a three-dimensional Fourier synthesis at 5.5Å resolution, obtained by X-ray analysis. *Nature* **185**:416–422.

Stout, G. H., and L. H. Jensen 1989. *X-ray Structure Determination: A Practical Guide.* 2nd edition. John Wiley and Sons, New York.

Structures of Proteins in Solution: Nuclear Magnetic Resonance

Bax, A. 1989. Two-dimensional NMR and protein structure. *Ann. Rev. Biochem.* **58**:223–256.

Clore, G. M., and A. M. Gronenborn 1991. Structures of larger proteins in solution: three and four dimensional heteronuclear NMR spectroscopy. *Science* **252**:1390–1398.

McIntosh, L. P., and F. W. Dahlquist 1990. Biosynthetic incorporation of ^{15}N and ^{13}C for assignment and interpretation of nuclear magnetic resonance spectra of proteins. *Quart. Rev. Biophys.* **23**:1–38. Other reviews on NMR and protein structure appear in the same volume.

Reid, B. R. 1987. Sequence-specific assignments and their use in NMR studies of DNA structure. *Quart. Rev. Biophys.* **20**:1–34.

Wüthrich, K. 1986. *NMR of Proteins and Nucleic Acids.* John Wiley & Sons, New York.

Wüthrich, K. 1989. The development of nuclear magnetic resonance spectroscopy as a technique for protein structure determination. *Acc. Chem. Research* **22**:36–44.

Wüthrich, K., et al. 1987. Nuclear magnetic resonance — from molecule to man. *Quart. Rev. Biophys.* **19**:3–237.

Suggestions for further reading on protein structure and function are listed at the end of Appendix 2.

2

How Proteins Recognize Each Other

Mutual recognition of proteins is crucially important in health and disease. The clarification of its stereochemical mechanisms has been one of protein crystallography's greatest achievements. Recognition is mediated mainly by two families of proteins: those that are like immunoglobulins and those that are like the proteins of the major histocompatibility complex. Immuno-globulins are either anchored in the cell membranes of lymphocytes or are secreted by them and circulate freely in the lymph and the blood; immunoglobulin-like domains also form part of many cell surface receptors. Proteins of the histocompatibility complex always remain anchored in cell membranes, but histocompatibility complex-like domains have also been found in soluble proteins. (See Chapter 9.)

This chapter first outlines the structures of immunoglobulins and the stereochemical basis of their specificity. It shows how immunoglobulins recognize the small protein lysozyme and how hybridoma and recombinant DNA technology, combined with X-ray crystallography, led to the design and production of a human-ized rat antibody that is effective against T-cell leukemia.

A quite different mechanism of recognition will then be seen to operate in immunity mediated by cells that carry proteins of the major histocompatibility complex on their surface. These proteins interact with two other proteins: the T-cell receptor and a protein called CD4; both contain immunoglobulin-like domains. CD4 has become a focus of medical interest since its discovery as the chief receptor molecule for the human immunodeficiency virus.

Neurons recognize their target cells by membrane-bound proteins, called neural cell adhesion molecules. They also contain immunoglobulin-like domains whose detailed structure can so far only be guessed. The activities of many proteolytic enzymes are controlled by specific inhibitors with surfaces that are complementary to those of the enzymes. Many inhibitors contain peptide loops with conformations that mimic the transition states of the enzymes' substrate.

We used to believe that the folding of nascent polypeptide chains to their specific three-dimensional structures is a spontaneous process both *in vivo* and *in vitro*, but the former has turned out to be much faster than the latter, because cells contain proteins called *chaperones* that accelerate the folding. One such chaperone will be shown to consist of two immunoglobulin-like domains capable of recognizing and embracing their targets.

Stereochemistry of Immune Reactions

A combination of chemical and X-ray analyses have revealed what immunoglobulins look like and how they work, while gene technology has provided the tools for making them to measure.

All immunoglobulins are made up of equal numbers of light and heavy chains. Each of the light ones contains about 220 and each of the heavy ones about 450–575 amino acid residues. The chains are divided into structurally similar domains of about 100 residues, each consisting of two layers of β-strands, usually stabilized by one internal disulphide bridge (Fig. 2.1). The light chains contain two and the heavy chains either four or five such domains. The joins between certain of the domains are flexible. The join between the variable and the first constant domain is referred to as the elbow, and that between the first constant domain and the remaining constant ones as the hinge, although "shoulder" would be a more descriptive term. The hinge is susceptible to proteolytic attack, yielding two fragments, the N-terminal one, called F_{ab} for antigen binding, and the C-terminal one, called F_c for complement-activating (Fig. 2.2).

Immunological specificity resides in the N-terminal domains, which contain polypeptide segments that are variable in the number and sequence of amino acid residues. These "hypervariable" segments form exposed loops, whose precise structure is determined by their amino acid sequence. Each of the chains contains three such hypervariable loops, also referred to as *complementarity-determining regions* or CDRs. They are the antigen combining sites. The invariant segments of the N-terminal domains form an almost rigid hydrogen-bonded framework held together by an

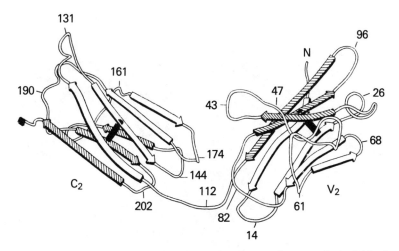

Figure 2.1. Mouse immunoglobulin light chain. Characteristic fold of the polypeptide chain in immunoglobulin domains. Arrows indicate directions of the chains in the two-layer pleated β-sheets. The black bars represent disulphide bridges. C: constant; V: variable domain. The numbers mark amino-acid residues in sequence from the amino-end, marked N. (Reproduced by permission, from Schiffer et al., 1973)

internal disulphide bridge. Immunological specificity is therefore provided by a variable pattern of amino acid residues mounted on an almost rigid frame (Fig. 2.3).

It looked at first as though the hypervariable loops could take up an almost infinite variety of different conformations in antibodies of different specificity, but this has turned out not to be true. By comparing the limited number of known antibody structures with the huge number of known antibody sequences, Chothia and Lesk (1987) found evidence in favor of only a limited number of conformations in five of the six loops. These so-called canonical structures can be recognized by certain key residues in specific positions along the sequence. Using these residues as their criteria, Chothia and Lesk were able to predict the correct conformations of the hypervariable loops in an F_{ab} fragment before its structure was solved by X-ray analysis (Chothia et al, 1987). By now all 70 V_H genes coding for residues 1–94 of the human heavy chains have been sequenced. In all but three, the hypervariable loops H_1 and H_2 can be predicted to have one of the few canonical structures, with varied sidechains decorating their surfaces. Most structural diversity rests in loop H_3.

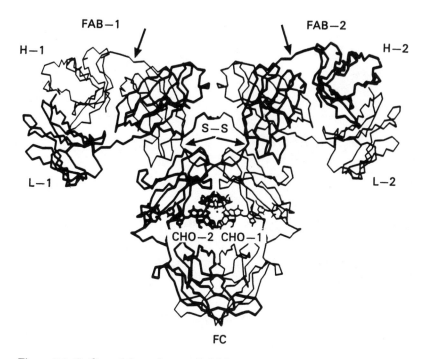

Figure 2.2. Outline of the polypeptide fold in an entire IgG immunoglobulin.
The heavy lines mark the α-carbon skeleton of the heavy chains, the
thin lines those of the light chains. FAB: antigen-binding segment.
FC: complement-activating segment; H: variable heavy chain domain;
L: variable light chain domain; The thin lines at the center of FC
represent carbohydrates, marked CHO. The links between the chains at
the center represent disulphide bridges. The angles at the elbows and at
the hinge are variable. The upper arrows point to the elbows, the lower
ones to the hinges. (Courtesy of Dr. Alan Edmundson)

    This information came from the solution of the structures of
several F_{ab} fragments, either alone or in combination with haptens,
which are small molecules that bind to the CDRs. The next step was
to find out how an antibody recognizes, and combines with, a
foreign protein. Does the antibody present a flexible surface that
adapts itself to that of the antigen, and vice versa, or is the antibody
"born" with a surface that is rigid and exactly complementary to a
rigid face of the antigen? This hotly argued questions was an
extension of the age-old controversy between two opposing
schools of immunologists whose views Macfarlane Burnett once

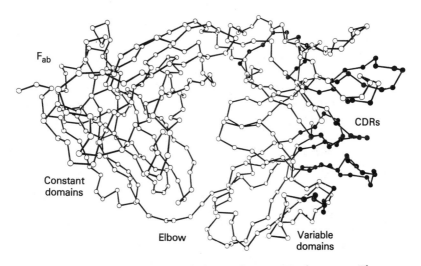

Figure 2.3. Outline of α-carbon skeleton of mouse F_{ab} fragment. The empty circles represent invariant residues, the full circles hypervariable residues in the complementarity-determining regions at the tip of the immunoglobulin.

compared to the alternatives of the rich man who has his suit made to measure and the poor man who buys it off the shelf.

Experiments using peptide fragments as antigens showed that these are often recognized by antibodies against the intact protein, indicating that antigenic activity resides in certain continuous segments of polypeptide chain on the protein surface. Such antigenic determinants or epitopes were widely believed to provide antigens with the flexibility needed to adapt themselves to the surface of the antibody, but in three different complexes of F_{ab}s with hen egg-white lysozyme, the structures of the bound proteins hardly differ from those of the free ones (Amit et al., 1986; Davies and Padlan, 1990; Bhat et al., 1990). Antibodies that do not actually cover the active site of lysozyme do not impair its catalytic activity, which is further evidence against any significant structural differences between the free and the antibody-bound enzyme. Similarly, it has been found that antibodies against foot and mouth disease virus combine with the unchanged structure of its coat proteins, which explains why a mutation resulting in the replacement of a single amino acid residue on the viral surface may suffice for the virus's escape from antibodies directed against its wild type. Solution of a complex of F_{ab} with influenza virus neuraminidase at first seemed to favor the rich man's way, but this conclusion is uncertain, because it was based on a partially refined structure of the complex, and it was reached without knowing the structure of the free F_{ab} (Tulip et al., 1989).

The three antilysozyme F_{ab}s combine with three quite separate surfaces of the lysozyme molecule, and between them they cover over 40% of its surface area (Fig. 2.4). The antibody and the lysozyme molecules each have a surface area of 680–780Å^2 in contact with their partner; these areas contain 75–110 atom pairs in contact, 10–23 hydrogen bonds, and up to three salt bridges. The contact areas of lysozyme are made up of two or more separate segments of polypeptide chain rather than of a single continuous one. The interacting surfaces are closely complementary and relatively flat: the interface of one complex contains no water, another only a single occluded water molecule, and the third two such molecules. These water molecules form hydrogen-bonded

Figure 2.4. A composite of the variable domains of three different monoclonal antibodies, HyHEL-5, D1-3, and HyHEL-10, bound to different surface regions of lysozyme at the center. The interacting surfaces have been separated for clarity. Together they cover over 40% of the lysozyme surface. (Reproduced, by permission, from Davies and Padlan, 1990)

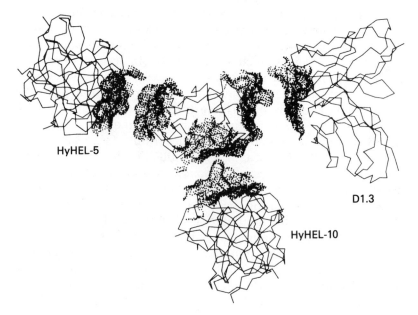

bridges between antibody and antigen. The structural differences between crystalline lysozyme alone and lysozyme combined with F_{ab} are small. Typically the root mean square displacement of the α-carbons is below 0.5Å and the biggest displacement of any single α-carbon is less than 2Å. Occasionally sidechains move or flip over (Davies and Padlan, 1990). The structures of the three lysozyme-F_{ab} complexes indicate that an animal immunized against a protein makes many different antibodies against it. Between them, these antibodies would cover the protein's entire surface in many different ways because, contrary to some current views, there is no part of that surface that is not antigenic. Nor is there any evidence that antigenic activity resides principally in flexible loops on the protein surface.

All six CDRs of the F_{ab}s are in contact with the lysozyme. In addition, there are some contacts between the lysozyme and the invariant frame at the base of the CDRs. The elbow bends of the free F_{ab} and the three F_{ab}– lysozyme complexes show widely different angles, but these are caused by different crystal packing rather than by interactions with the antigen. Does the antibody adapt its structure to fit the antigen? Solution of the structure of one of the free F_{ab}s excluded major differences between free and bound antibodies, but the structure lacked the resolution needed to discover minor changes. Poljak and his collaborators have been able to obtain the necessary resolution by crystallizing a complex (F_V) of only the variable domains of the light and heavy chains (V_L and V_H) with and without lysozyme. Superposition of the free V_L domain on the lysozyme-bound V_L, or of the free V_H domain on the lysozyme-bound V_H gave a root mean square deviation of the α-carbon atoms of only 0.37Å, showing adaptation to the antigen within each domain to be minimal. On the other hand, when the free V_L and the lysozyme-bound V_L domains were superimposed, the root mean square deviation between the α-carbon positions of the free and the lysozyme-bound V_H domain was 0.99Å. This represented a small movement of the V_L and V_H domains relative to each other that does bring them into closer contact with the antigen by a mechanism of induced fit (Bhat et al., 1990). Returning to Macfarlane Burnett's parable, antigens do take their antibodies off the peg, but *small* alterations to both improve their mutual fit.

Mice develop a syndrome related to human systemic lupus erythematosus. Spleen cells from one such mouse that were fused with nonsecreting myeloma cells produced a monoclonal antibody that bound single-stranded DNA instead of peptide or protein. In this antibody, the place of the usual hapten-binding cavity between the CDRs of the heavy and light chains is taken by a long irregular groove, lined with aromatic amino acids that may become sandwiched between the bases of the DNA, and with lysines and arginines designed to bind to its phosphates. This structure illustrates most strikingly the immunoglobulins' remarkable adaptive capacity (Herron et al., 1987).

Organisms make antibodies against antibodies, known as anti-idiotopes. An anti-idiotope against an antibody to an antigen was believed to resemble the antigen itself, but this expectation was disproved by X-ray analysis of a complex of an antilysozyme F_{ab} with its anti-idiotope F_{ab}. Their two F_{ab}s are aligned with their long axes roughly in parallel (Fig. 2.5). All but one of their twelve CDRs interlock, and one CDR also makes contact with the invariant frame of its partner F_{ab}. The hypervariable faces of the two antibodies are displaced relative to each other, so that one of them binds mainly to the V_L domain and to the groove between the V_L and V_H domains of the other. Since both faces are concave, this arrangement maximizes the contacts between them. There is no significant difference between the structures of the antilysozyme F_{ab} when it is bound to lysozyme or to the anti-idiotope. There is a topological resemblance, but no chemical resemblance, between the face of lysozyme and that of the anti-idiotope in contact with the antilysozyme F_{ab}. The two proteins have achieved topological complementarity to the same antibodies in quite different ways (Bentley et al., 1989).

Knowledge of the structure of antibodies bound to anti-idiotopes is medically important, because anti-idiotopes have been used successfully in the treatment of certain lymphomas and leukemias (Miller et al., 1982), and they may also find uses in the treatment of autoimmune diseases.

Engineering a Human Antibody Against T-cells

Natural antibody diversity arises from the haphazard shuffling of genes and from somatic mutations, but genetic engineering combined with X-ray analysis now permits us to make antibodies to measure. Hybridoma technology has been most successful in small rodents and has led to the manufacture of a wide range of monoclonal antibodies for clinical diagnostic tests. For treating patients, on the other hand, human antibodies are needed, because any foreign ones are liable to cause adverse immune reactions, but monoclonal antibodies cannot be produced in humans. L. Riechmann, M. R. Clark, H. Waldmann, and G. Winter have overcome this difficulty by taking a human myeloma protein of known structure and altering

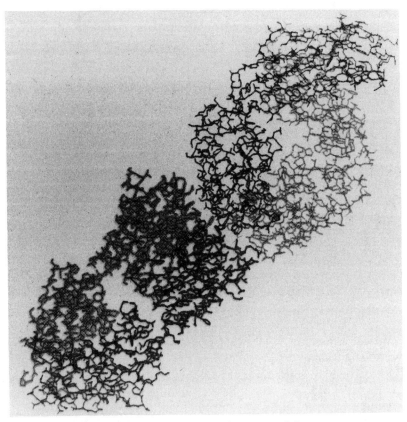

Figure 2.5. Structure of a complex of the F_{ab} fragment of the mouse monoclonal antibody D1-3, top right, with the F_{ab} fragment of its anti-idiotope E225, bottom left, showing the contact between the complementarity-determining regions of the two immunoglobulins. (Reproduced, by permission, from Bentley et al., 1990)

parts of its amino acid sequence by genetic engineering, so as to provide it with a high affinity for human T-cells as well as the ability to trigger the destruction of these cells. They started with a rat antibody that kills human T-cells. This had been used in immunosuppression of graft-versus-host reactions that cause rejection of transplanted organs and in the treatment of T-cell leukemia. It is an immunoglobulin of the IgG type. IgGs are dimers, made up of one pair of light chains and one pair of heavy chains (Fig. 2.2).

To convert the rat anti–T-cell antibody into a human one, Waldman, Winter, and their colleagues grafted the oligonucleotides that code for the CDRs of the rat antibody onto the genes coding for the heavy and light chains of two different human myeloma proteins of known structure. They then grafted onto the gene for the heavy chain the nucleotides coding for the C-terminal domains of another myeloma protein that was known to activate complement. Finally, they joined the heavy and light chains to a single IgG molecule (Fig. 2.2). Having accomplished that elaborate feat of genetic engineering, they found their antibody to be inactive.

Winter then asked if the rigid frame of the N-terminal domains of the human IgG onto which he had grafted the rat CDRs, differed from the frame of the rat IgG in any way that would interfere with recognition by the T-cells. He found that in the heavy chain of the rat IgG the side chain of an internal phenylalanine provided a prop that kept one of the CDRs in position. In the human frame that phenylalanine was replaced by a serine that left a large gap between the supporting frame and the CDR. Winter therefore replaced that serine codon in the gene for the human heavy chain by the codon for a phenylalanine. That single replacement activated the antibody. It could never have been found without knowledge of the atomic structure (Riechmann et al., 1988).

The Cambridge group made sufficient antibody for the treatment of two patients with terminal non-Hodgkin lymphoma for whom other treatments had failed. Doses of 1–20 mg of the engineered antibody were given for up to 43 days. In both patients lymphoma cells were cleared from the blood and bone marrow, and the enlarged spleen shrank to normal size. One patient had lymphadenopathy that was also resolved. Normal hematopoiesis was restored, partially in one patient and completely in the other. Both could be discharged from the hospital and could lead a normal life. One patient was free of symptoms for 15 months, but died from multiple secondary tumors two years after treatment had been begun; the other was still alive and well two years later (Hale et al., 1988). Further clinical trials now await manufacture of the IgGs on a larger scale. They may have wider clinical applications (for example, as immunosuppressive agents after tissue transplantation and for the treatment of some autoimmune diseases).

Another Approach to Autoimmune Disease: Structures of the Major Human Histocompatibility Complex and of the T-cell Receptor
Immune reactions are of two kinds: humoral, through secreted antibodies, and cell-mediated. The first step in cell-mediated immunity consists in phagocytosis of the foreign invader by a

macrophage. The macrophage splits the invading proteins into small fragments and attaches these fragments to proteins of the major histocompatibility complex (MHC), also called human leukocyte antigen (HLA). These proteins migrate to the surface of the macrophage where the foreign fragments carried by them are recognized by T-cells. Once a T-cell has recognized a foreign antigen presented to it by the histocompatibility protein, it is activated to grow into a clone of cells armed to recognize any protein of which that fragment is a part. If such proteins show up on the surface of another cell, the T-cells destroy it (Marrack and Kapler, 1986). There are two main classes of HLA molecules: class I is recognized by cytotoxic (killer) T-cells that can destroy virus-infected, tumor, or foreign cells, while class II is recognized by helper T-cells that regulate the production and secretion of immunoglobulins by B-cells. Bjorkman, Wiley, and their colleagues at Harvard have determined the structure of the class I human leukocyte antigen HLA-A2 that was originally found to be associated with acute lymphatic leukemia in children. The structure shows how the histocompatibility complex binds peptide fragments and presents them to the T-cell receptor. The histocompatibility molecule is made up of two different pairs of domains. One pair is similar to the domains of the immunoglobulins. The other pair forms the active site which looks like a vice designed to clamp the antigen. This location of the active site is confirmed by the positions of the amino acid residues that differ in different serological types of class I MHCs. They are all clustered along the jaws and the bottom lining of the vice. Moreover, X-ray analysis shows a molecule between its jaws that is not part of MHC and is presumably a foreign antigen bound there (Figs. 2.6 – 2.8) (Bjorkman et al., 1987; Saper et al., 1991).

Garrett, Wiley, and others have determined the structure of a second class I human histocompatibility antigen, an allele called HLA-AW68, which differs from HLA-A2 at 13 amino acid positions. The two structures are closely similar: the root mean square difference in α-carbon positions is only 0.45Å. Ten of the 13 amino-acid replacements are in the lining of the specificity pocket, on either the α-helices or the pleated sheet (Fig. 2.8). They alter both the shape and the charges of part of the lining. Beneath one of the helices the substitution His $74\rightarrow$Asp opens a deep pocket, designed to fit a cationic sidechain. Alterations in the lining of that pocket could add to the variability provided by the lining of the vice. The entire vice and the pocket are filled with electron density, presumably representing an unresolved mixture of peptides which

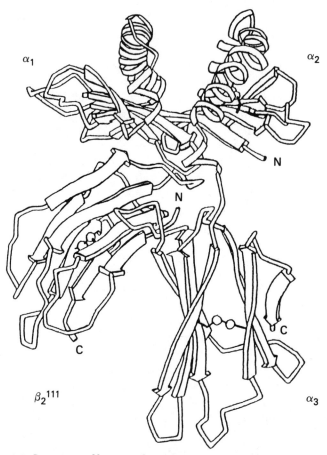

α_1

α_2

N

N

C

C

β_2^{111}

α_3

Figure 2.6. Structure of human class I histocompatibility antigen HLA-A2. General view of the molecule after removal of the transmembrane anchor attached to the α-chain by digestion with papain. This leaves the heavy α-chain with three domains: α_1 and α_2 (each made up of an α-helix and a pleated sheet) and α_3 (which has a structure similar to the domains of immunoglobulins shown in Figure 2.1). The fold of the β_2-chain, known as β_2-microglobulin, is also immunoglobulin-like. (Reproduced, by permission, from Bjorkmann et al., 1987)

the HLA-AW 68 presents to the T-cell receptor (Garrett et al., 1989; Bjorkman and Parham, 1990).

Peptides of various length split from the nucleoprotein of the influenza virus were found to bind to class I MHC molecules with dissociation constants ranging from $30\mu M^{-1}$ to $300\mu M^{-1}$. A nonapeptide bound best (Cerundolo et al., 1991). Preference for nonapeptides

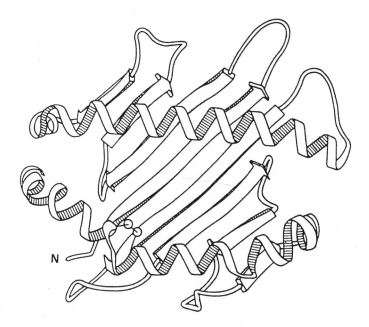

N

*Figure 2.7. The HLA molecule seen from the top of Figure 2.6. The two
α-helices form the jaws of a vice. The pleated β-sheets form its base.
(Reproduced, by permission, from Bjorkman et al., 1987)*

*Figure 2.8. Stereoview of the vice, seen from the top, showing the 10
amino acid replacements in HLA-AW68 compared to HLA-A2.
(Reproduced, by permission, from Garrett et al., 1989)*

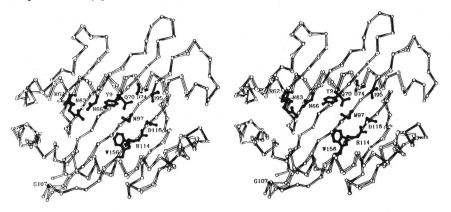

is confirmed and explained by a well-resolved X-ray structure of a class I MHC associated with the autoimmune disease ankylosing spondilitis. Clamped in its vice is a mixture of "self" nonapeptides with an almost fully extended β-structure. The charged termini of the peptides bind to conserved complementary charges of the protein; the spacing apart of these charges evidently determines the preference for nonapeptides. In addition, four sidechains of the peptide dip into specificity pockets of the protein. For example, one pocket with hydrophobic walls ends with a glutamate, clearly designed to hold an arginine or lysine sidechain. Another pocket is lined with aromatic sidechains that would attract either aromatic or aliphatic sidechains of the peptide. Recognition of only four of any of a nonapeptide's sidechains makes the MHC versatile and likely to bind at least one of the many different nonapeptides split from a foreign protein, which it can then present to a T-cell receptor. Four of the peptide's sidechains reach into the solvent or hug the surface of the vice and should therefore be recognized by that receptor (Madden et al., 1991).

There is as yet no crystal structure of any class II MHC molecule, but the patterns of variable and constant amino acid residues in members of the two classes are so similar as to leave little doubt that their antigen-binding sites have similar structures (Samraovi et al., 1988).

Different alleles of MHC are associated with a variety of allergic and autoimmune diseases such as ankylosing spondilitis, rheumatoid arthritis, insulin-dependent diabetes, and coeliac disease. Understanding of the stereochemical mechanism of cell-mediated immunity may be the first step toward the development of rational treatments of such diseases (Batchelor and McMichael, 1987).

No one has yet solved the structure of any T-cell receptor, but a great deal can be inferred from the amino acid sequences of these receptors. They are made up of two different polypeptide chains: α and β. Each chain has a sequence that begins with two immunoglobulin-like domains which are followed by a hinge region, a hydrophobic transmembrane helix, and a cytoplasmic domain. The sequence of the N-terminal domain is variable, with hypervariable segments in positions equivalent to those of immunoglobulins, while the sequence of the second domain is constant. The sequences look as though the extracellular portion of the T-cell receptor had a structure similar to that of a single domain of an immunoglobulin molecule. These purely analytical data inspired two independent groups to propose similar plausible models of a variable T-cell receptor bound to a less variable MHC molecule. They face each other so that the receptor's CDRs 1 and 2 recognize and bind to the helical jaws of the MHC vice, while the CDRs 3 recognize the peptide antigen clamped between them (Fig. 2.9) (Davis and Bjorkman, 1988; Chothia et al., 1988).

The CD4 Receptor and Its Binding to HIV

Peptide antigens bound to class II HLAs are anchored to the surface of a macrophage or a B-cell. T-cells recognize them with the help of two immunoglobulin-like molecules: the T-cell antigen receptor

Figure 2.9. Hypothetical model of a T-cell receptor in contact with a protein of the major histocompatibility complex (MHC 2) with its bound antigen. Top: variable domain of the F_{ab} fragment J539. Bottom: α_1 and α_2 domains of HLA-A2 with its bound peptide antigen P. Since this picture was drawn, Madden et al. (1991) have found a peptide antigen bound to MHC 1 to be not α-helical, but extended. This now seems the more likely conformation of MHC-bound antigens. (Reproduced, by permission, from Davis and Bjorkman, 1988)

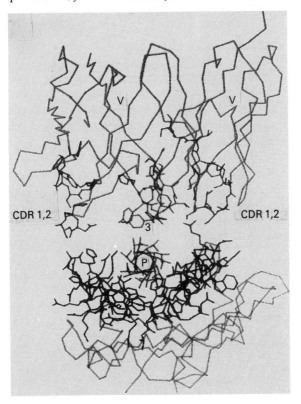

just described and a protein called CD4. Both are anchored to the surface of the T-cell (Staunton et al., 1989; Giranda et al., 1990; Rayment et al., 1982; Liddington et al., 1990) (Fig. 2.10). The sequence of the 433 amino acid residues in its single polypeptide chain indicates that its extracellular part is made up of four immunoglobulin-like domains linked in tandem. These are anchored to a hydrophobic transmembrane helix and a small cytoplasmic domain. Crystals of the entire CD4 molecule gave poor X-ray diffraction patterns, but crystals of only the two N-terminal domains, which constitute the part that recognizes HLA, have

Figure 2.10. Diagrammatic representation of the T-cell receptor (TCR), combined with the glycoprotein CD4, interacting with the major histocompatibility complex (MHC II) and its bound protein antigen (PA) in the contact between a macrophage (top) and a T-cell (bottom). The horizontal lines indicate the respective cell membranes. (Reproduced, by permission, from Travers, 1990)

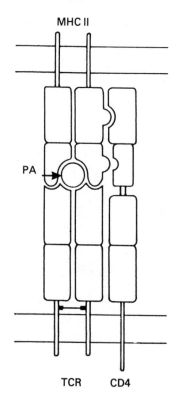

yielded to X-ray analysis (Ryu et al., 1990; Wang et al., 1990). Their structure is indeed immunoglobulin-like: superposition of the 53 common residues of the N-terminal domain of CD4 on the variable light chain domain of a Bence-Jones protein gave a root mean square deviation of main chain atomic position of only 0.83Å. In that domain, a disulphide bridge links β-strands in neighboring pleated sheets as in immunoglobulins, while in the second domain the disulphide bridge links β-strands within the same sheet, a feature that had not been seen before. The N-terminal domain presents four external loops, like the three CDRs of the immuno-globulins (Fig. 2.11). The strength of the interaction between CD4 and HLA is affected by amino acid substitutions in both domains of CD4, which suggests that the two molecules make contact all along their lengths.

Figure 2.11. Polypeptide fold of the two immunoglobulin-like domains of CD4. (Reproduced, by permission, from Ryu et al., 1990)

CD4 is also the receptor for the human immunodeficiency virus (HIV) and may mediate its entry into the host cell. It has a high affinity for the envelope protein of HIV, gp160, and also for its split product, gp120. Most amino acid substitutions in CD4 that alter that affinity lie on its second external loop (Fig. 2.12). Their distribution indicates that the area of contact between the two proteins measures at least 12 x 25Å, an area of a size similar to the contact area between a single immunoglobulin chain and its antigen. The CD4 loop may

Figure 2.12. Structure of the human lymphocyte protein CD4 with residues whose replacement affects its binding to the HIV surface protein gp120. Circles represent exposed residues and squares buried ones. (Reproduced, by permission, from Wang et al., 1990)

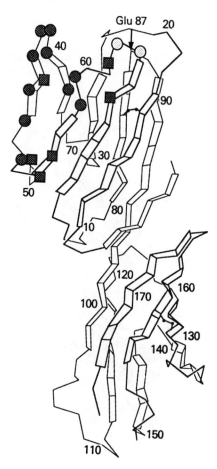

fit into a canyon-like receptor lined with invariant amino acids below the surface of the viral envelope, similar to the receptor sites of the rhino and influenza viruses (see Chapter 8). The structure of CD4 provides some clues to the design of inhibitors of HIV binding. Structural knowledge of gp120 and of the gp120–CD4 complex would be even better, but so far all attempts to crystallize either of these have failed.

HIV-infected cells tend to fuse with as yet uninfected cells that carry CD4 on their surface. The fusion is set in motion by the binding of HIV envelope proteins of the infected cell to CD4 of the uninfected one. Chimpanzee CD4 also binds human HIV and gp120, but the infected cells do not fuse and infected chimpanzees do not develop AIDS. Chimpanzee CD4 has a glycine in place of glutamate 87, which lies at the tip of the third external loop (Figs. 2.11 and 2.12). When this same substitution is made in human CD4, gp120 still binds and the human cells become infected, but they do not fuse. Taken together with the effects on infectivity of amino acid substitutions in the second loop, these findings suggest that binding of the second loop to the viral envelope is sufficient for infection, but binding of both the second and the third loops is needed for fusion (Barbacid, 1987; Robey and Axel, 1990; Camerini and Seed, 1990). The N-terminal domain of the CD2 antigen, which makes T-lymphocytes and accesory cells adhere to each other, also has an immunoglobulin-like fold (Driscoll et al., 1991).

Neural Cell Adhesion Molecules, Heat-Shock Proteins,
and Chaperones
Once nature has evolved an efficient fold of the polypeptide chain, it tends to use it over and over again for related, and sometimes even for quite different, purposes. Some of these folds can be recognized without X-ray analysis, because they exhibit characteristic amino acid sequence patterns. This has led to the recognition of a wide range of immunoglobulin- and MHC-like proteins.

In 1955 the great biologist Johannes Holtfreter discovered that dissociated embryonic cells of amphibia reaggregate and form tissue-like structures, because like cells joined up and unlike ones segregated (Townes and Holtfreter, 1955). Recently such recognition was found to be mediated by families of specific proteins on cell surfaces. One of these, the family of neural cell adhesion molecules, makes embryonic neurons adhere to other neurons and to muscle cells (Cunningham et al., 1987).

Cunningham and others have determined the amino acid sequences of three different neural cell adhesion molecules derived from a single chicken gene by alternative splicing of its messenger RNA. A leader sequence is followed by five homologous, extracellular domains of about a hundred residues each, then by a transmembrane helix, and finally, in two of the three molecules, by a cytoplasmic domain. The five homologous domains exhibit immunoglobulin-like sequences, including the two cysteines spaced 50–56 residues apart that form the characteristic disulphide bridge between adjoining pleated sheets (Fig. 2.1), a tryptophan whose indol ring leans against that bridge, four other invariant residues, and the alternation of hydrophobic residues on the inside with hydrophilic ones on the outside of the two-layered pleated sheet. Cunningham and others suggest that like neural cell adhesion molecules on the surfaces of adjacent cells recognize each other by, say, the first domain of one molecule adhering to the fourth domain of its neighbor and vice versa.

Organisms ranging from bacteria to mammals produce a special class of ATP-binding proteins in response to heat and other forms of stress. The functions of these heat-shock proteins remained obscure until Hugh Pelham suggested in 1986 that they combine with unfolded polypeptide chains, either preventing or reversing their aggregation, helping them to refold to their native structure, and finally releasing them, with hydrolysis of ATP (Pelham 1986). His ideas have since been confirmed by many experiments (Ostermann et al., 1989; Flynn et al., 1989; Craig and Gross 1991). They have shown similar proteins to be present also in unstressed cells, where they keep nascent proteins unfolded to help their passage through the endoplasmic reticulum and also perform other important functions. Perhaps because they protect unfolded proteins from intracellular proteinases, heat-shock proteins are also called *chaperones* or, more awkward, *chaperonines*. It will be interesting to find out whether they also play a role in human trauma and fever.

The most abundant heat-shock proteins have molecular weights of about 70 kD. In addition to the functions that I have already mentioned, these proteins strip coated vesicles of their protein coats. Coated vesicles perform endocytosis, i.e., they transfer outside matter into cells; their coats are made of the protein clathrin. Once inside the cells, they are shorn of their clathrin coats and fuse into larger vesicles called endosomes that process the imported matter (Stryer, 1988).

Flaherty and others isolated a bovine 70 kD heat-shock protein with clathrin-binding specificity. They subjected it to proteolysis,

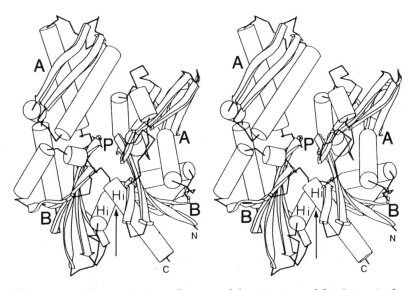

Figure 2.13. Schematic stereo diagram of the structure of the C-terminal fragment of the heat-shock protein HSC70, with α-helices represented as cylinders and β-strands as arrows, showing the location of the A and B domains. P indicates the ATP-binding site. Hi shows the helices enclosing the hinge between the domains, indicated by the arrow. (Reproduced, by permission, from Flaherty et al., 1990)

which yielded two fragments, an N-terminal 44 kD fragment that bound clathrin and a C-terminal 70 kD fragment that hydrolyzed ATP. They crystallized the latter and determined its structure (Fig. 2.13) (Flaherty et al., 1990). It looks like a purse, designed to pocket nature's energy coin ATP. Each side of the purse is lined by two domains. The two B-domains are made of seven-stranded β-pleated sheets flanked by three α-helices, and their structures are almost superimposable, while the structures of the two A domains are distinct. The electron density map also revealed an unexpected Mg-ATP complex buried deep in the pocket, with nearby aspartates and glutamates ready to catalyze its hydrolysis.

To the authors' surprise, they found the structure of the B-domains to be similar to that of the corresponding domains of yeast hexokinase, the enzyme that catalyzes the phosphorylation of glucose by ATP. The hexokinase purse is wide open in the absence of substrate. Glucose binds to one of its walls and ATP to the other, but they cannot meet as long as the purse is open. Closure brings them together and initiates catalysis (Fersht, 1985). Daniel Koshland

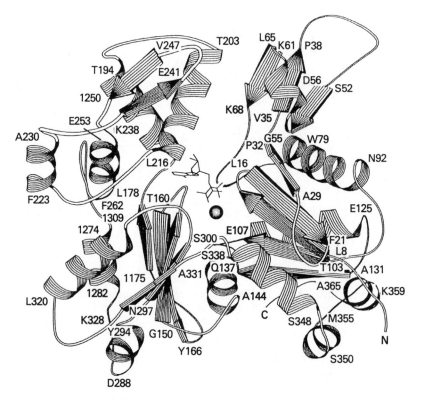

Figure 2.14. Schematic diagram of the structure of rabbit muscle actin in a similar orientation as HSC70 in Figure 2.13, showing ATP and Mg²⁺ in the active site. (Reproduced, by permission, from Kabsch et al., 1990)

inferred a mechanism of this kind many years before the structure was solved, and called it "induced fit" (Koshland, 1959).

An even greater surprise came with the subsequent publication of the structure of rabbit muscle actin; it was found to be identical to that of the heat-shock protein's ATP-binding domains even though their amino acid sequences show no homology (Fig. 2.14) (Kabsch et al., 1990). Actin polymerizes into long filaments whose interaction with myosin causes muscle to contract. Structural identity of two functionally and chemically so unrelated proteins must be an extreme example of nature's use of the same polypeptide fold for different purposes.

There is no X-ray structure as yet of the C-terminal clathrin-binding domain of the 70 kD heat-shock protein, but its amino acid sequence shows weak homology with the antigen-binding domains of MHC proteins (Fig. 2.7) (Rippmann et al., 1991). This suggests an intriguing mechanism for the ATP-linked binding and release of clathrin and other proteins by the heat-shock protein. The opening and closing of the hexokinase purse operates mainly by a relative rotation and shift of the two helices marked Hi (for hinge) in Figure 2.13 (Lesk and Chothia, 1984). Seeing that those helices lie at the C-terminus of the ATP-binding domain, they must be in contact with the clathrin-binding domain. Closing of the purse with hydrolysis of ATP could therefore transmit a leverage to the MHC-like clathrin-binding domain, causing it to open its helical jaws and release the clathrin chain that was clamped between them. Flynn and others suggest that the heptapeptide-binding site of another 70 kD heat-shock protein has a structure similar to the MHC binding site (Flynn et al., 1991).

Jacob has compared nature's way of evolving new structures to the tinker's art of making new gadgets from bits of old ones. The architecture of the heat-shock protein is a good example of such tinkering on the molecular scale (Jacob, 1982).

An entirely different chaperone that is neither a heat-shock protein nor an ATP-ase has been found in a strain of *E. coli* that is associated with the kidney disease pyelonephritis. These bacteria attach themselves to their host cells with pili made up of several proteins whose assembly is mediated by a chaperone called *PapD*. Holmgren and Brändén found it to be made up of two immunoglobulin-like domains inclined to each other like the blades of a boomerang (Fig. 2.15). They enclose a concave patch of exposed valine and isoleucine sidechains, flanked by three glutamates on one side and two arginines and a lysine on the other. Holmgren and Brändén suggest that this region of the molecule chaperones the assembly of the pili proteins. No other immunoglobulin-like protein has so far been found in bacteria (Holmgren and Brändén, 1989).

Serine Proteinases and Their Inhibitors

The activities of many proteolytic enzymes are controlled by protein inhibitors that recognize and combine with their active sites. While antigens select from millions of different antibodies those that happen to have a roughly complementary surface, complementarity between enzymes and their natural inhibitors has evolved more exactly, so that they can bind to each other more tightly. In at least one instance X-ray analysis has proven that the complex of

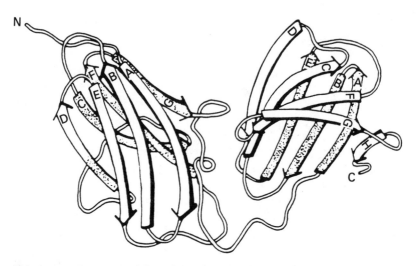

Figure 2.15. Diagram of the chaperone protein PapD, showing two domains, each with β-strands in two layers. The connections between the strands are the same as in the immunoglobulin domains shown in Figure 2.1, but the domains are turned upside down with respect to those in Figure 2.1. (Reproduced, by permission, from Holmgren and Bränden, 1989)

enzyme and inhibitor mimics the transition state of the substrate. In this section I shall concentrate on the serine proteinases and their inhibitors, because they are among the most important physiologically and also the best studied. They constitute a large family from which I have selected the pancreatic proteinases and the pancreatic trypsin inhibitor; leukocyte elastase and α_1-antiproteinase; the plasminogen activator inhibitor; thrombin, antithrombin, and hirudin.

The structure of the first serine proteinase, chymotrypsin, was solved in 1967 by David Blow with Brian Matthews, Paul Sigler, and Richard Henderson, after surmounting a succession of seemingly insuperable hurdles in a seven-year marathon (Matthews et al., 1967). They solved all except the structure of the active site, partly because an error in the chemical sequence had mistakenly identified the crucial aspartate as an asparagine. When that structure was solved two years later, it provided the key to our understanding of the catalytic mechanism of a great variety of enzymes, including those of the blood-clotting cascade (Blow et al., 1969). The structure strikingly confirmed Linus Pauling's (1948) prediction:

I think that enzymes are molecules that are complementary in structure to the activated complexes of the reactions that they catalyze, that is, the molecular configuration that is intermediate between the reacting substances and the products of reaction.

Serine proteinases split peptide and ester bonds. Figure 2.16 shows the typical tertiary structure of mammalian serine proteinases. Their active site lies in a cleft between two domains of mainly β-strands and contains a serine, a histidine, and a buried aspartate linked by hydrogen bonds (Figure 2.17). The aspartate is ionized and polarizes the histidine, which in turn polarizes the serine hydroxyl and renders its oxygen strongly nucleophilic. The catalytic mechanism, which is one of the best studied in biochemistry, is believed to work as follows (Formula 2.1). When a peptide substrate

Figure 2.16. Folding of the polypeptide chain in the coat protein of the Sindbis virus (see Chapter 8). The fold is typical of a large class of serine proteinases, including the pancreatic enzymes trypsin, chymotrypsin, and elastase. S215, H141, and D164 mark the catalytic residues in the active site cleft. In the pancreatic enzymes their equivalent would be serine 195, histidine 57, and aspartate 102. In the coat protein the C-terminal tryptophan, marked C, blocks the active site, shown in detail in Figure 2.17, page 66. (Reproduced, by permission, from Choi et al., 1991)

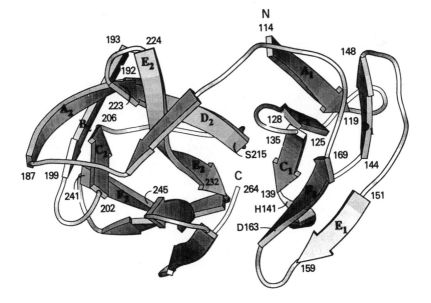

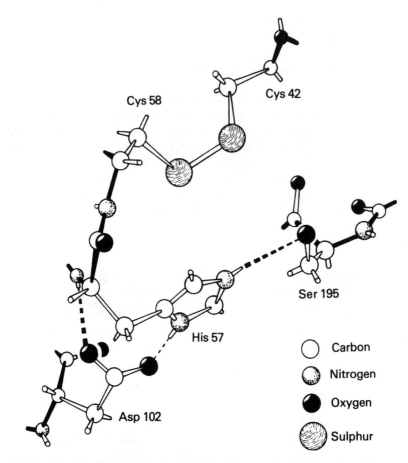

Figure 2.17. Catalytic triad of serine, histidine, and aspartate found in serine proteinases. The broken lines represent hydrogen bonds. (Reproduced, by permission, from the original diagram of Blow, Birktoft and Hartley, 1969)

Formula 2.1. (opposite) The hydrolysis of a peptide. (A): Acylation. A tetrahedral transition state is formed, in which the peptide bond is cleaved. The amine component then rapidly diffuses away, leaving an acyl-enzyme intermediate. (B): Deacylation. The acyl-enzyme intermediate is hydrolized by water. Note that deacylation is essentially the reverse of acylation, with water in the role of the amine component of the original substrate.

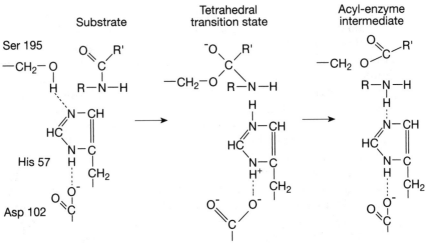

A

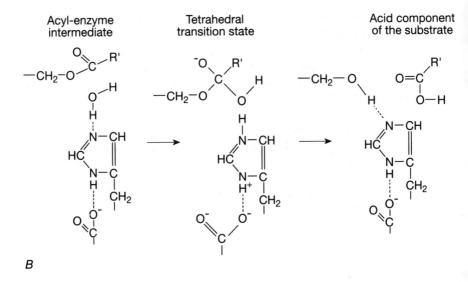

B

binds, its carbonyl carbon faces the serine oxygen; its carbonyl oxygen accepts hydrogen bonds from two main chain NH groups; and its peptide nitrogen faces N_ε of the neighboring histidine. The first step in catalysis consists in the formation of a transition state intermediate in which the serine oxygen forms a loose bond with the carbonyl carbon. This carbon is now bonded to four atoms at the corners of a tetrahedron instead of three atoms at the corners of a triangle. The carbonyl oxygen becomes negatively charged. Its charge is stabilized by the hydrogen bonds donated by the two main chain NH groups.

The second step in catalysis consists in rupturing the peptide bond and forming an acyl intermediate. In this step the serine hydroxyl donates a proton to the neighboring histidine and forms a covalent bond with the substrate's carbonyl carbon. The histidine donates a proton to the peptide NH. These two events break the peptide bond and liberate the C-terminal polypeptide.

The third step consists in hydrolysis of the acyl intermediate and liberation of the N-terminal polypeptide. This is accomplished by the binding of a water molecule to the histidine. The water molecule donates one of its protons to the histidine and becomes an OH^-. The negatively charged oxygen attacks the substrates' carbonyl carbon, breaks the acyl bond, and converts the carbonyl into a carboxylate (Fersht, 1985; Stryer, 1988).

Serine proteinases are specific for the amino acid residue that precedes the scissile bond. Specificity is determined by a pocket next to the active site. Chymotrypsin, which cleaves the bonds following aromatic residues, has a deep, mainly hydrophobic pocket. Trypsin, which cleaves the bonds following basic residues, has an aspartate at the base of its pocket. Elastase, which is specific for short aliphatic sidechains, has a shallow, mainly hydrophobic pocket.

The pancreatic serine proteinases are synthesized as inactive precursors that are activated in the digestive tract. Chymotrypsinogen, the precursor of chymotrypsin, is a protein of 245 amino acid residues. Trypsin activates it by splitting the bond between arginine 15 and isoleucine 16. The α-amino group of this isoleucine then buries itself in a pocket where it forms a strong salt bridge with the carboxylate of an aspartate. In the absence of that bond, the active site and the specificity pocket are loose and incompletely formed. This explains the inactivity of the zymogen.

The activation step is irreversible, but the activities of many serine proteinases are kept in check by protein inhibitors that mimic the transition states of their substrate and fit to the protein-ases very tightly. For example, the pancreatic trypsin inhibitor is a protein of only 56 amino acid residues that fits into trypsin like an electric plug into its socket, with a dissociation constant of 10^{-13}. The surface area buried between the two proteins (1400Å^2) is of the same order as that between a typical antigen and an antibody. Tight binding is due to formation of strong hydrogen bonds between main chain amides of proteinase and inhibitor, both in and near the active site cleft. The inhibitor loop that enters the cleft contains a lysine followed by an isoleucine. The lysine's basic sidechain fits into the enzyme's specificity pocket where it forms a salt bridge with a buried aspartate. The carbonyl carbon connecting the lysine to the isoleucine enters the active site and combines with the serine hydroxyl to form a near-analogue to the tetrahedral transition state of a peptide substrate; yet hydrolysis of the inhibitor's peptide bond takes several weeks, because the two residues forming that bond are tightly clamped between two nearby cystine bridges, so that inhibition of the enzyme is almost irreversible. Except for its mobile sidechains, trypsin and its inhibitors combine as rigid bodies, so that little binding energy is dissipated by loss of entropy (Fig. 2.18).

Pancreatic is now known to be a misnomer for this inhibitor. It occurs mainly in mast cells that happen to be present also in the pancreas. The inhibitor is most abundant in the lungs where its function is unknown. On the other hand, the health of the lungs seems to depend on proper function of a serine proteinase inhib-itor in blood plasma named, also misleadingly, α_1-antitrypsin, or better, α_1-antiproteinase. This inhibits neutrophil elastase, the chief matrix-cleaving proteinase secreted by activated leukocytes. It belongs to a large family of inhibitors known as serpins, of which it is the best-studied example. It is a glycoprotein of 394 amino acid residues in a single chain. Its specificity for elastase is determined by a methionine. The structure of the intact α_1-antiproteinase is still unknown, but is can be inferred from that of its inactive homologue egg albumin (Stein et al., 1990). This is made up of two layers of β-pleated sheets partly covered by α-helices. Protruding from its compact structure is an α-helical loop, like a lamp filament. Its counterpart in α_1-antiproteinase is a peptide loop that contains the active methionine, but to fit into the active site cleft of leukocyte elastase, the loop in α_1-antiproteinase should be extended rather than helical (Fig. 2.19A).

Unlike the lysine–isoleucine bond in the pancreatic trypsin inhibitor, the bond between the active methionine and the serine

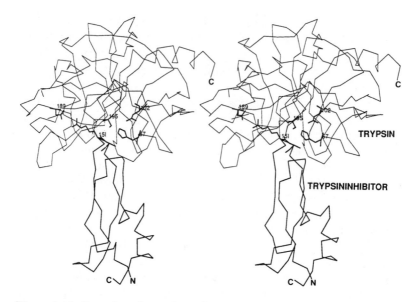

Figure 2.18. Complex of trypsin and pancreatic trypsin inhibitor. The stereodiagram show only the α-carbon positions, except in the active site where the sidechains of the catalytic triad serine 195, histidine 57, and aspartate 102 are drawn. Also drawn is aspartate 189, at the base of the specificity pocket that is filled with the sidechain of the inhibitor's lysine 42. (Courtesy of Professor R. Huber)

that follows it in α_1-antiproteinase is easily cleaved. The structure of that cleaved form has been solved (Löbermann et al., 1984). It shows cleavage to be followed by a drastic and irreversible change of structure. The active methionine and the serine are separated from each other by about 70Å, and the active site loop has become incorporated in the six-stranded pleated sheet of the inhibitor in the form of an additional β-strand.

Homozygous carriers of certain mutant α1-antiproteinase genes suffer from a predisposition to emphysema of the lung. For example, 4% of Northern Europeans carry the mutation glutamate 342→lysine which reduces secretion of the antiproteinase to less than one sixth of normal. Homozygotes for this mutation are predisposed also to cirrhosis of the liver, because the mutant protein aggregates in hepatocytes after its synthesis and clogs them with inclusion bodies. Smoking contributes to emphysema of the lung because tobacco smoke oxidizes the reactive methionine sidechain to sulfoxide, which is a misfit in the specificity pocket of leukocyte elastase, and the α1-antiproteinase therefore fails to inhibit it. The same happens during inflammation when neutrophils release oxygen radicals that inactivate α_1-antiproteinase, thereby freeing the elastase secreted by them to attack surrounding tissues.

Burial of the active loop between the β-strands explains why substitution of glutamate 342 by lysine causes α_1-antiproteinase to polymerize and aggregate in the liver. In Fig. 2.19B this glutamate lies at the tip of the black arrow that marks the hinge of the flexible loop. Lomas, Evans and Carrell (1992) show that the substitution allows the flexible loop of one α_1-antiproteinase molecule to insert itself between the two β-strands that would flank the flexible loop of another α_1-antiproteinase molecule (if the loop were buried). The second molecule's flexible loop would insert itself into a third, and so on until a long chain would form. A rise in temperature from 37° to 41°C accelerates polymerization because it promotes the opening of the gap between the two β-strands. Hence prevention of fever would help to prevent the severe liver disease to which babies and small children afflicted by this disorder are prone.

Blood clotting is triggered by a cascade of enzymatic reactions that ends with the proteolytic activation of prothrombin to thrombin. This is a serine proteinase that cleaves four arginine-glycine bonds in soluble fibrinogen and converts it to insoluble fibrin. Thrombin has a three-dimensional structure similar to that of the pancreatic serine proteinases (Bode et al., 1989). (See Fig. 2.16.) It is inhibited by the serpin antithrombin which is homologous to α_1-antiproteinase, except for the substitution of the active methionine by arginine and a 25-residue N-terminal extension that binds heparin. Without heparin, antithrombin is inactive.

Some years ago, a boy from Pittsburgh suffered from severe bleeding even though no ill-function of any of his clotting enzymes could be found. Eventually his bleeding was discovered to be due to a mutation in his α_1-antiproteinase gene, which had replaced its active methionine by arginine, thus making his α_1-antiproteinase masquerade as antithrombin (Owen et al., 1985).

Serpins can take up two alternative conformations: an active, apparently stressed one in which part of their inhibitory peptide loop is exposed as in egg albumin, and an inactive, relaxed form in which that loop has snapped back and buried itself at the other end of the pleated sheet, with its C-terminal arm exposed and its N-terminal arm wedged into that sheet as in the cleaved α_1-antiproteinase.

The structure of the intact relaxed and inactive form has been revealed by an X-ray analysis of the human plasminogen activator

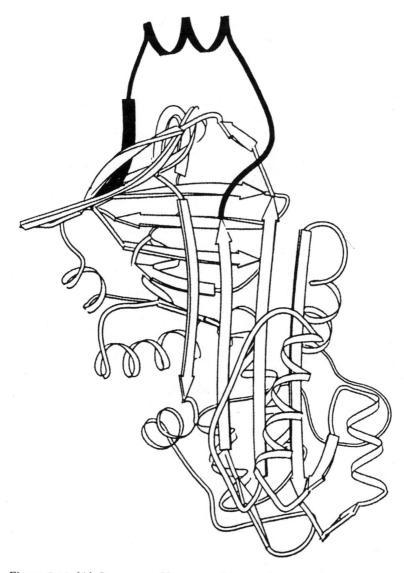

Figure 2.19. (A) Structure of hen egg-white ovalbumin. The black
α-helical loop at the top occupies the position in the amino acid
sequence equivalent to that of the active site loop in α₁-antiproteinase.
The active methionine would occupy the top left-hand corner of the
α-helix, but in α₁-antiproteinase the loop is probably extended rather
than helical.

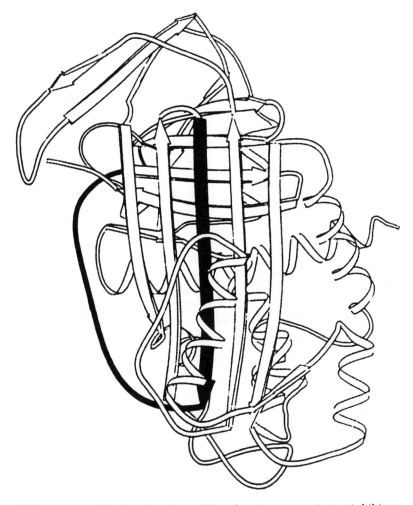

(B) *Structure of the inactive form of the plasminogen activator inhibitor. The active site loop, drawn in black, is folded back into the molecule, forming an additional β-strand midway between the other four β-strands. (Courtesy of Dr. Arne Strand and Dr. Elizabeth Goldsmith)*

inhibitor-1 (PAI-1 for short) (Mottonen et al., 1992). (See Fig. 2.19 B.) Plasminogen is the precursor of plasmin, the serine proteinase that splits and dissolves fibrin clots. Its activator is a protein of 72 kD now manufactured by recombinant DNA technology for the treatment of coronary thromboses. The inhibitor of that activator, PAI-1, also inhibits urokinase, a widely used bacterial activator of plasminogen. *In vivo*, PAI-1 is maintained in its active, metastable state by combination with the 50 kD protein vitronectin. When freed from vitronectin, the active peptide loop is folded back to its stable, inert position.

Antithrombin and PAI-1 seem to have evolved a spring-loaded safety catch that makes them revert to their latent, stable, inactive forms unless the catch is kept in position by another molecule: antithrombin by heparin, PAI by vitronectin, any serpin by the proteinase that it is designed to inhibit. Only when that safety catch is in position is the active site loop of these serpins exposed and ready for action; otherwise it snaps back and hides inside the protein. α_1-antitrypsin may exist in an equilibrium between the two forms in plasma even when it is free. Spring-loaded protein molecules represent a remarkable new biological control mechanism.

Hirudin is a powerful anticoagulant secreted by the medicinal leech. It is a peptide of 65 amino acid residues that forms a complex with thrombin quite unlike the complex with antithrombin. The N-terminal amino group of hirudin enters the active site of thrombin and attaches itself with a hydrogen bond to the OH of the catalytic serine. Its other residues fill the active site cleft, wrap themselves around the enzyme, and cling to it with 10 salt bridges and 22 hydrogen bonds (Rydel et al., 1990).

Conclusions

All molecular recognition depends on steric and electrostatic complementarity of the combining surfaces. When proteins recognize each other and form stable complexes, the surfaces buried between them cover large areas. Janin and Chothia (1990) have made a comparative study of the combining surfaces of antigens and antibodies, and of proteinases and their inhibitors in published structures of their complexes. They found the buried surface areas to be of roughly the same size in both: $1600(\pm350)\text{Å}^2$. The dissociation constants of antigen–antibody complexes vary from 10^5 to 10^{10} M^{-1}, but those of proteinases from their inhibitors can be even smaller. For example, the dissociation constant of the complex between the F_{ab} fragment DI.3 and lysozyme shown in Figure 2.4 is $3.4 \times 10^9 M^{-1}$ at 20°C, compared to 10^{13} M^{-1} of trypsin from the pancreatic trypsin inhibitor. Not all the proteinase–proteinase inhibitor complexes are as stable. The hydrophobic effect, dispersion forces, and hydrogen bonds contribute to their binding energies in varying degrees. (For the relative strength of these bonds, see Appendix 2.) Janin and Chothia found between 8 and 13 hydrogen bonds between combining surfaces, some of them formed by one ionizable group with another either neutral or oppositely charged group. They found no systematic difference between the mobilities of surface segments of proteins that combine specifically with other proteins, and the average mobility of surface segments of all proteins of known structure.

The mutant α_1-antiproteinase that afflicted the boy from Pittsburgh illustrates the decisive importance of a single charge in determining specificity. The methionine at the active site of normal α_1-antiproteinase prevents its combination with thrombin, because entry of the neutral methionine sidechain into the specificity pocket of thrombin would bury the negatively charged carboxylate of the aspartate at the base of the pocket. Buried uncompensated and unhydrated charges are known to destabilize proteins so much that they unfold. On the other hand, when the mutant α1-antiproteinase combines with thrombin, the positively charged guanidinium group of the arginine at its active site is compensated by the negatively charged aspartate at the base of thrombin's specificity pocket, which stabilizes the complex. Hydrogen bonds stabilize protein–protein complexes, while burial of uncompensated hydrogen bond donors or acceptors destabilizes them.

Further Reading

Bjorkman, P. J., and P. Parham 1990. Structure, function and diversity of class I major histocompatibility complex molecules. *Ann. Rev. Biochem.* **59**:253–288.

Bode, W., and R. Huber 1991. Ligand binding: Proteinase and proteinase inhibitor interactions. *Opinions in Structural Biology* **1**:45–52.

Bode, W., and R. Huber 1992. Natural protein proteinase inhibitors and their interaction with proteinases. *Eur. J. Biochem.* **204**:433–51.

Carrell, R. W., K. S. Aulak, and M. C. Owen 1989. The molecular pathology of the serpins. *Mol. Biol. Med.* **6**:35–42.

Carrell, R. W., and D. R. Boswell 1986. Serpins: The superfamily of plasma serine proteinase inhibitors. In *Proteinase Inhibitors*, A. Barrett, and G. Salveson, eds. Elsevier Science, Amsterdam, Ch. 12, pp. 403–422.

Chothia, C., and A. M. Lesk 1987. Canonical structures for the hypervariable regions of immunoglobulins. *J.Mol.Bol.*, **196**:901–917.

Chothia, C., A. M. Lesk, A. Tramontano, M. Levitt, S. J. Smith-Gill, G. Air, S. Sheriff, E. A. Padlan, D. Davies, W. R. Tulip, P. M. Colman, S. Spinelli, P. M. Alzari and R. J. Poljak 1989. Conformations of immunoglobulin hypervariable regions. *Nature* **342**:877–883.

Davies, D. R., E. A. Padlan, S. Sheriff 1990. Antibody-antigen complexes. *Ann. Rev. Biochem.* **59:**439–474.

Davies, D. R., S. Sheriff, and E. A. Padlan 1988. Antibody-antigen complexes. *J. Biol. Chem.* **263:**10541–10544.

Gething, M. J., and J. Sambrook 1992. Protein folding in the cell. *Nature,* **355:**33-45. This is a review on heat-shock proteins, chaperones and other enzymes involved in protein folding.

Stryer, L. 1988. *Biochemistry.* 3rd edition. W. H. Freeman, New York.

Tsomides, T. J., and H. N. Eisen 1991. Antigenic structures recognized by cytotoxic T lymphocytes. *J. Biol. Chem.* **266:**3357–60.

Watson, J. D., N. H. Hopkins, J. W. Roberts, J. A. Steitz, and A. M. Weiner 1987. The generation of immunological specificity. In *Molecular Biology of the Gene,* vol. 2, pp. 832–899. Benjamin/Cummings, Menlo Park, CA.

3

How Proteins Recognize Genes in Health and Disease

Transcription of genes into messenger RNA is controlled by switches attached to specific DNA loci, sometimes near to, and sometimes far from, the genes to be transcribed. These switches are proteins whose amino acid sequence is determined by other genes that may be located elsewhere in the genome. Proteins whose combination with DNA turns transcription on are called *activators*, and those whose combination with DNA turns transcription off are called *repressors*. In prokaryotes, the specific locus on the DNA with which a switch combines is called the *operator*, and the collection of genes whose transcription it controls is called the *operon*. The affinity of the activator or repressor for its operator may be controlled by a metabolite, a hormone, a kinase, a proteinase, or a variety of other chemical devices acting either as inducers or inhibitors. Genetic activators and repressors constitute the main controls of growth, differentiation, and oncogenesis. Therein lies their medical importance.

The first lead to the nature of the genetic control of transcription came in 1950 from an abstruse, seemingly irrelevant phenomenon that had puzzled biologists in the first half of this century. This was lysogeny, a name given to latent viral infection of bacteria. Certain strains of *E. coli* infected with bacteriophage λ undergo lysis with massive release of phage, while other, equally infected strains multiply for many generations with only sporadic liberation of phage by a few of the bacteria. These lysogenic strains are immune to infection with the same phage. Some virologists believed that the sporadic liberation of virus arose from bacterial genes that had escaped from their normal coordination and control; others regarded the virus as a disordered element due to degeneration from the normal bacterium's perfect Platonic order.

Max Delbrück, the skeptical founder of phage genetics, dismissed lysogeny as an artifact due to sloppy experimentation, but André Lwoff, a biologist at the Pasteur Institute in Paris, thought otherwise and refused to believe that lysis was stochastically determined. He said in his Nobel lecture: "My soul is not statistical; I am a protozoologist and I like to look at things rather than calculate probabilities."

François Jacob (1988) writes:

> Lwoff tackled the problem in his usual original manner. Within a few weeks he had transformed it. He did not want to study whole populations of bacteria like everybody else, but single individuals. With a micromanipulator he fished out of his culture one infected, lysogenic bacterium after another and followed their behavior under the microscope. He found an astonishing phenomenon: all the descendants of the ancestor that had once encountered the virus preserved the memory of that event. Even after hundreds, nay thousands of generations, every descendant still had the potential to produce the same virus as his ancestor.

In each individual the hidden virus persisted in an inactive, dormant state which Lwoff christened *prophage*, on the grounds that a name makes things real. Lwoff decided to try and wake the virus up. Here is his own account.

> Negative experiments piled up, until after months and months of despair, it was decided to irradiate the bacteria with ultraviolet light. This was not rational at all, for ultraviolet radiation kills bacteria and bacteriophages, and on a strictly local basis the idea still looks illogical in retrospect. Anyhow, a suspension of lysogenic bacilli was put under the UV lamp for a few seconds.
>
> The Service de Physiologie Microbienne is located in an attic, just under the roof of the Pasteur Institute, with no proper insulation. The thermometer sometimes rises in a manner that leaves no conclusion other than that the temperature is high. After irradiation, I collapsed in an armchair, in sweat, despair, and hope. Fifteen minutes later, Evelyne Ritz, my technician, entered the room and said: "Sir, I am growing normally." After another quarter of an hour, she came again and reported simply that she was normal. After fifteen more minutes, she was still growing. I was very hot and more desperate than ever. Now sixty minutes had elapsed since irradiation; Evelyne entered the room again and said very quietly, in her soft voice: "Sir, I am entirely lysed." So she was: the bacteria had disappeared! As far as I can remember, this was the greatest thrill — molecular thrill — of my scientific career. (Lwoff, 1966)

Lwoff's discovery transformed people's views about the nature of viruses and their relationship with their hosts. If a single bacterium can transmit infection with a latent virus through thousands of generations, then the viral genome must be integrated in the bacterial one and duplicated every time a bacterium divides. Its activation by ultraviolet light suggested that expression of the viral gene remained repressed until radiation damage inactivated the repression.

By a spectacular leap of the imagination Lwoff's younger colleagues François Jacob and Jacques Monod recognized in 1960 that the molecular mechanism for switching on expression of the viral genes by irradiation with ultraviolet light is the same as the one they had found to switch on the synthesis of lactose-metabolizing enzymes in *E. coli* when the inducer, lactose, was added to the culture medium. That idea led them to formulate the concept of the operon whose recognition as the universal mechanism for controlling gene expression in prokaryotes won them, together with Lwoff, the Nobel Prize for Physiology or Medicine for 1965 (Jacob and Monod, 1961).

For several years after the discovery of genes for the λ and lac repressors their chemical nature remained obscure, because their concentration is normally very low: a single wild type *E. coli* cell contains only about 10 copies of the lac repressor. In 1966 Gilbert and Müller-Hill at Harvard engineered a strain of *E. coli* that produced a 2,000 times larger quantity of repressor, which they isolated by equilibrium dialysis against a "gratuitous" inducer, that is, a lactose analogue that is not cleaved by β-galactosidase. They found that the repressor is a tetramer made up of four identical subunits of 37 kD, each binding one molecule of inducer (Gilbert and Müller-Hill, 1966). The tetramer binds to the DNA operator with the astonishingly low dissocation constant of 10^{-13} M^{-1}. In the following year Ptashne isolated the λ-repressor and found it to be a dimer of two identical proteins of 26 kD (Ptashne, 1967). The operators to which these repressors bind are short sequences of nucleotide bases that read the same forwards and backwards, like the word *madam*. They are called *palindromes* from the Greek *palindromos* (running back again).

Complete explanation of Lwoff's induction of phage growth by ultraviolet light had to wait much longer. Such light breaks DNA chains. Their rupture induces *SOS repair*, which is a system of enzymes that repairs breaks in double-stranded DNA. The first of these enzymes is *rec A*, which catalyzes recombination of DNA strands, but also induces the dimeric λ-repressor to split apart. Its monomers then dissociate from the operator, which sets in motion a chain of events that results in the expression of the phage genes.

After Jacob and Monod's initial discoveries, chemical and structural studies of transcriptional control in prokaryotes progressed rapidly, but at the same time evidence for such controls in eukaryotes continued to remain purely genetic. For example, in yeast the product of a gene called GAL4 was found to induce transcription of the genes for galactose-metabolizing enzymes, but not until 1987 was that gene product isolated and characterized (Johnston, 1987).

In the mid-1980s the almost simultaneous discovery of two different DNA-binding protein motifs—the zinc fingers and the homeodomains—finally transformed the outlook for an understanding of genetic control in eukaryotes. This chapter describes their structures and the manner of their binding to specific DNA sequences, including the remarkable similarity between the structure of the λ-repressor of *E. coli* with its DNA operator and that of the homeodomain of a fruit fly with its DNA homeobox. The structures of homeodomains have now assumed medical importance because translocation of their genes from one human chromosome to another has been recognized as a frequent cause of children's leukemia. It is a poetic vindication of Lwoff's persistent probing of a seemingly minor biological phenomenon as well as a striking manifestation of the unity of nature on the molecular scale.

c-Jun and c-Fos are two other transcription regulators that can become oncogenic, as are certain zinc fingers. The binding of these regulatory proteins to DNA appears to be strengthened by a succession of peptide loops containing the sequence serine or threonine-proline-X-X-; there are as yet no X-ray structures of this peptide sequence nor of its complex with DNA, but chemical experiments indicate that the structure of that complex is similar to that of the complex of an anticancer drug known to bind to the minor groove of double-helical DNA. This peptide motif may also be involved in oncogenesis.

Eukaryotic DNA-binding motifs are often part of much larger proteins that change their structures in response to chemical stimuli in ways that we do not yet understand. Their complete structures and the mechanisms of their response remain to be explored.

Zinc Fingers and Hormone Receptors

In 1985 Miller, McLachlan, and Klug discovered that TFIIIA, a DNA-binding zinc protein that controls the transcription of the 5S RNA genes in Xenopus oocytes consists of a tandem repeat of 9 homologous domains of 30 residues, each containing pairs of closely spaced cysteines and histidines (Fig. 3.1) (Miller et al., 1985).

1 (M G E K A L P V V Y K R) 12

```
1   Y I Ⓒ S F A D Ⓒ G A A Ⓨ N K N W K Ⓛ Q * A Ⓗ L C * K Ⓗ    37
2   T G E K * P Ⓕ P Ⓒ K E E G Ⓒ E K G Ⓕ T S L H H Ⓛ T * R Ⓗ S L * T Ⓗ    67
3   T G E K * N Ⓕ T Ⓒ D S D G Ⓒ D L R Ⓕ T T K A N M K * K Ⓗ F N R F Ⓗ    98
4   N I K I C V Ⓨ V Ⓒ H F E N Ⓒ G K A Ⓕ K K H N Q Ⓛ K * V Ⓗ Q F * S Ⓗ    129
5   T Q Q L * P Ⓨ E Ⓒ P H E G Ⓒ D K R Ⓕ S L P S R Ⓛ K * R Ⓗ E K * V Ⓗ    159
6   A G - - * - Ⓨ P Ⓒ K K D D S Ⓒ S Ⓕ V G K T W T Ⓛ Y L K Ⓗ V A E C Ⓗ    188
7   Q D - - * L A V Ⓒ - - D V Ⓒ N R K Ⓕ R H K D Y Ⓛ R * D Ⓗ Q K * T Ⓗ    214
8   E K E R T V Ⓨ L Ⓒ P R D G Ⓒ D R S Ⓨ T T A F N Ⓛ R * S Ⓗ I Q S F Ⓗ    246
9   E E Q R * P Ⓕ V Ⓒ E H A G Ⓒ G K C Ⓕ A M K K S Ⓛ E * R Ⓗ S V * V Ⓗ    276
```

277
(D P E K R K L * K E K C P R P K R S L A S R L T G Y I P P K S E K [N] A 311
S V S G T E K T D S L V K N K P S G T E T [N] G S L V L D K L T I Q) 344

Figure 3.1. Amino acid sequence of Xenopus zinc finger. Homologous residues are circled. For meaning of single letter amino acid code, see Figure A2.1 on page 278. (Reproduced, by permission, from Miller et al., 1985)

The authors suggested that each quartet of cysteines and histidines binds one zinc ion and that the peptide loop connecting the pair of cysteines to the pair of histidines formed a DNA-binding finger (Fig. 3.2). That zinc finger has become the most common eukaryotic DNA-binding motif discovered so far. Between 0.5 and 1% of the human genome codes for zinc finger motifs in perhaps hundreds or more different proteins. Zinc fingers have been found to play pivotal roles in the control of eukaryotic gene transcription. Intense efforts have therefore been made to determine their structures and those of their complexes with DNA by X-ray analysis and magnetic resonance spectroscopy (Klug and Rhodes, 1987).

In 1988 Berg proposed a structure for the zinc finger from comparisons of its amino acid sequence with those of metallo-enzymes of known structure (Berg, 1988) (Fig. 3.3). It is centered around a zinc ion that is tetrahedrally coordinated to the two cysteines and two histidines. Going from the amino to the carboxyl terminus, the structure is made up of a β-strand, followed by a hairpin bend that contains the two cysteines, then a second β-strand antiparallel to the first, a hairpin loop, and finally a three-turn α-helix with the zinc-bound histidines on successive turns. The space between the helix and the β-strand is filled with hydrophobic side chains. Berg also predicted that zinc fingers would bind to DNA with their α-helices fitting into its major groove, which would allow successive fingers to wrap around the double helix.

Figure 3.2. Miller, McLachlan, and Klug's interpretation of the zinc finger sequence shown in Figure 3.1. Ringed residues are conserved. Black dots mark proposed DNA-binding residues. (Reproduced, by permission, from Miller et al., 1985)

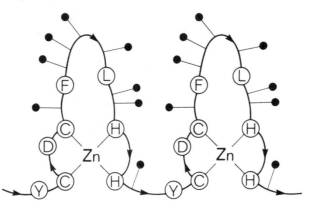

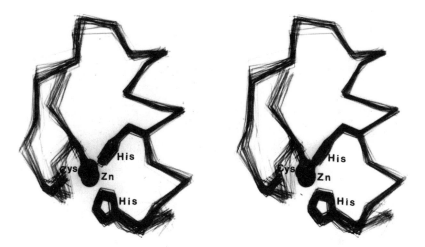

Figure 3.3. Course of the polypeptide chain of the zinc finger and zinc coordination predicted by Berg and then confirmed by NMR. (Reproduced by permission, from Lee et al., 1989)

Berg's predictions have been proved exactly right by the structures of single zinc fingers in solution determined by NMR and the structure of a crystalline three-zinc-finger–DNA complex determined by X-ray analysis (Párraga et al., 1988; Billeter et al., 1989; Lee et al., 1989; Pavletich and Pabo, 1991). The complex was obtained by crystallization of a 90 amino acid residue DNA-binding domain cleaved from a mouse protein and combined with a 10 base-pair double helical DNA.

Figure 3.3 shows the course of the polypeptide chain of a single finger and its coordination to the zinc ion (Lee et al., 1989). Figure 3.4A shows the amino acid sequence of the three mouse zinc finger domains that exhibit strong homology with those of the Xenopus zinc fingers. Figure 3.4B shows the base sequence of the double helical oligonucleotide to which the three mouse zinc fingers were bound. Figure 3.5 shows the detailed structure of the first of the three mouse zinc fingers with its α-helix to the left and its anti-parallel pleated sheet to the right of the tetrahedrally coordinated zinc ion. Figure 3.6 on page 85 is a diagrammatic view of the DNA complex. The α-helices of the three fingers do indeed fill the major groove and make equivalent contacts with successive trios of bases. Rotation of each finger by 96° around the DNA helix axis and a shift of about 10Å along that axis brings it into congruence with the next finger. The α-helices are inclined by 45° to the planes of the base pairs.

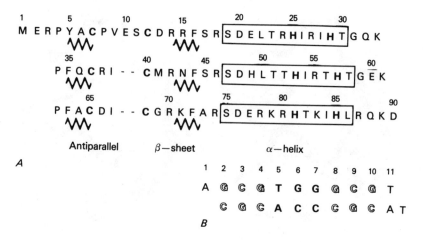

Figure 3.4. (A) Amino acid sequence of mouse zinc fingers. Bold C's and H's mark the cysteines and histidines coordinated to the zinc ions. (B) Base sequence of double-helical oligonucleotide to which the zinc fingers are bound. (Reproduced, by permission, from Pavletich and Pabo, 1991)

Figure 3.5. Detailed structure of one of the mouse zinc fingers. Note the hydrophobic side chains of phenylalanine 16 and leucine 22 that fill the space between the α-helix on the left and the β-strands on the right. Also note arginines 18, 24, and 27, and lysine 33 that extend from the α-helix toward the DNA. (Reproduced, by permission, from Pavletich and Pabo, 1991)

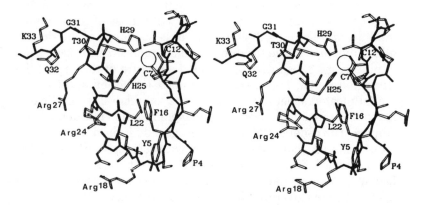

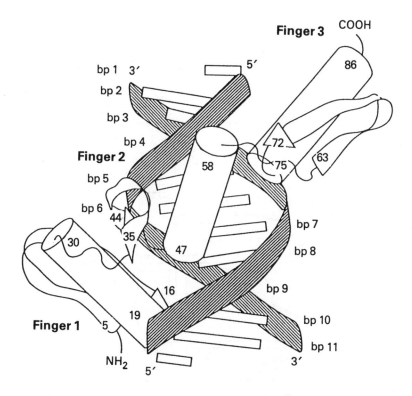

Figure 3.6. Schematic diagram showing the binding of the mouse zinc fingers to the double helical oligonucleotide. The cylinders represent α-helices. The numbers on the protein refer to the amino acid sequence in Figure 3.4A, the base pair numbers to the sequence shown in Figure 3.4B. (Reproduced, by permission, from Pavletich and Pabo, 1991).

Sidechains in each finger form hydrogen bonds with DNA phosphates and bases (Fig. 3.7). The arginines immediately preceding the amino-ends of each helix are hydrogen bonded to guanines. Each of these arginines is held in position by a hydrogen bond to an aspartate attached to the α-helix (Fig. 3.8, page 87). Two other arginines and a histidine in the α-helices also bind to guanines, while five other arginines and one serine form nonspecific hydrogen bonds with DNA phosphates. In each finger one of the histidines that is coordinated to zinc with its N_ε also donates a hydrogen bond to a phosphate oxygen with its N_δ. The hydrogen bonds between arginines and guanines mediate specific recognition;

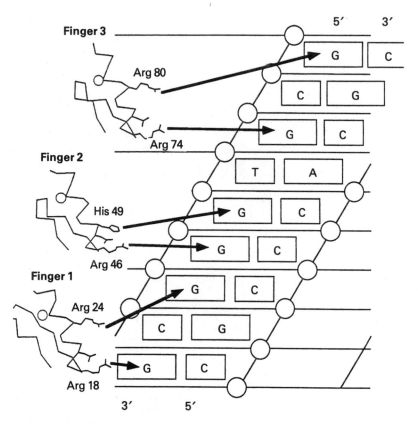

Figure 3.7. Hydrogen bonds from arginines and a histidine in the mouse zinc fingers to guanines in the DNA. (Reproduced, by permission, from Pavletich and Pabo, 1991)

Pavletich and Pabo suggest that their model could form the basis for synthetic zinc fingers designed to recognize other bases.

Zinc fingers are the first known DNA-binding proteins with a variable number of modular repeats that are geometrically adapted to bind to the repeats of the DNA double helix. The binding of the first finger would reduce the loss of rotational and translational entropy needed to bind the second finger and so on for the binding of successive fingers, which would make their binding cooperative, meaning that the binding energy of n fingers would be much larger than the sum of the binding energies of n individual fingers. Nevertheless the binding energies of zinc fingers to DNA and their specificities are much weaker than those of prokaryotic repressors.

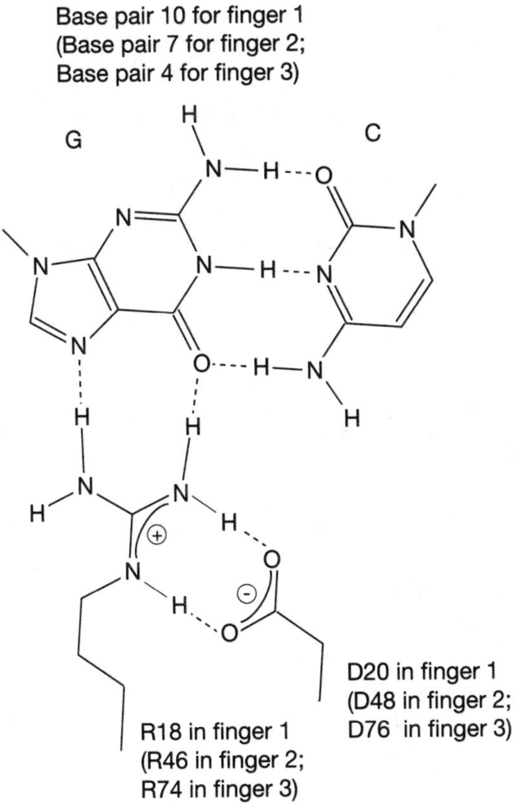

Base pair 10 for finger 1
(Base pair 7 for finger 2;
Base pair 4 for finger 3)

R18 in finger 1
(R46 in finger 2;
R74 in finger 3)

D20 in finger 1
(D48 in finger 2;
D76 in finger 3)

Figure 3.8. Details of hydrogen bonds between the ion pair of arginine and aspartate and the base pair of guanine and cytosine in the mouse zinc fingers. (Reproduced, by permission, from Pavletich and Pabo, 1991)

For example, the three-finger protein binds to its specific base sequence a hundred times more strongly than to random DNA, while the λ repressor of E. coli binds to its operator 10^7–10^8 times more strongly than to random DNA. The reason may lie in the greater flexibility of the zinc fingers which raises their loss of entropy on binding to DNA.

NMR methods have also been used to solve the structures to two medically important zinc fingers that had not yet been crystallized. They are the DNA-binding domains of the glucocorticoid and estrogen receptors. Glucocorticoids such as cortisol promote gluconeogenesis and glycogen synthesis, and they enhance the degradation of fat and protein, while estrogens are needed for the development of secondary female sex characteristics and the function of the ovarian cycle. Their sites of action are in the cell nucleus where they activate the synthesis of new proteins. After entering the cytoplasm of a cell, these hormones combine tightly with specific receptor proteins that diffuse to the cell nucleus, where they bind to specific DNA promoter sequences, called hormone response elements, located 5' to the genes to be activated (Fig. 3.9). Receptor binding to those elements initiates transcription of the genes into messenger RNA, followed by protein synthesis.

The DNA-binding domains of the receptor proteins are separate from their hormone-binding domains. They are rich in cysteines and contain bound zinc ions, reminiscent of the Xenopus transcription factor TFIIIA. The 70-odd-residue-long DNA-binding domain of the estrogen receptor contains two zinc ions, each tetrahedrally coordinated to four cysteines, but neither its amino acid sequence nor its structure is exactly homologous to the zinc fingers of Xenopus oocytes. The domain includes two mutually perpendicular α-helices, each followed by regions of extended chain. Between the helices lies a core of hydrophobic residues. Each zinc is bound to two cysteines at successive turns of an α-helix and to two cysteines that are part of the extended loops; these loops form two similar "zinc fingers," which have a structure distinctly different from that of the TFIIIA-type zinc fingers (Fig. 3.10) (Schwabe et al., 1990). Mutagenesis studies have shown that replacement of glutamate 25,

Figure 3.9. Nucleotide base sequence in the DNA of the response elements for the glucocorticoid (GRE), estrogen (ERE), and thyroid hormone (TRE) receptors. GRE differs from ERE in two base pairs at each half site, while TRE differs from ERE in the two half sites being closer together. The sequences have been symmetrized. (Reproduced, by permission, from Schwabe et al., 1990)

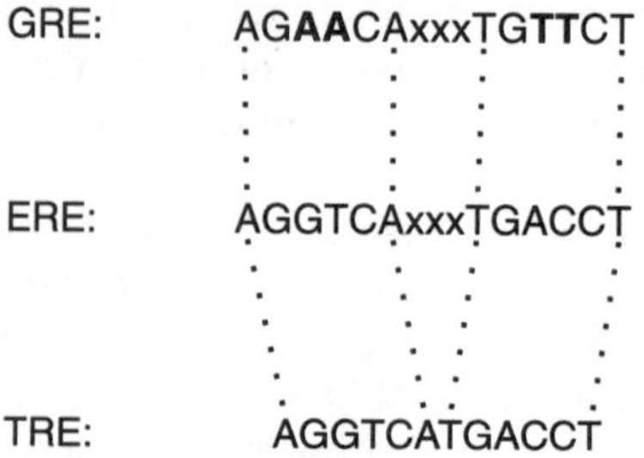

GRE: AGAACAxxxTGTTCT

ERE: AGGTCAxxxTGACCT

TRE: AGGTCATGACCT

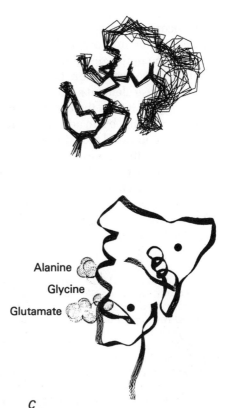

Figure 3.10. DNA-binding domain of the estrogen receptor. (A) Stereodrawing of an overlay of 15 independently calculated structures of the domain, selected for their favorable geometry and few violations of the constraints imposed by the NMR results. (B) The fold of the polypeptide chain with its two zinc atoms (black circles) and the sidechains in the hydrophobic core. (C) The same, seen at right angles to B with the sidechains of the three DNA-recognizing residues on the outside of one of the helices. (Reproduced, by permission, from Schwabe et al., 1990)

glycine 26, and alanine 29 at the N-terminus of the first α-helix by glycine, serine, and valine, respectively, changes the specificity of the receptor, so that it recognizes the glucocorticoid rather than the estrogen response element, which differs from the former by only two base pairs (Fig. 3.9) (Mader et al., 1989; Umesono and Evans, 1989).

The structure of the DNA-binding domain of the glucocorticoid receptor was also determined by NMR and found to be similar to that of the estrogen receptor (Härd et al., 1990). In solution, the domain is a monomer, but two domains bind cooperatively to the major glucocorticoid response element and form a dimer when bound. Luisi, Sigler, and others have determined the crystal structures of two DNA complexes of that dimer (Luisi et al., 1991). In one complex, solved at 4Å resolution, the two complementary hexanucleotides of the response element's palindrome were separated by three base pairs as in Figure 3.9. In the other complex, solved at 2.9Å solution, they were separated by an additional, fourth base pair. The DNA binding domain consisted of residues 440–525 of the glucocorticoid receptor (Fig. 3.11). Its structure in the crystalline dimeric DNA-complex is consistent with that found for the single free domain in solution. The dimer is held together by hydrogen bonds and by contacts between hydrophobic sidechains buried at the interface between the two C-terminal helices that face away from the DNA double helix (Fig. 3.12). The N-terminal α-helices that are part of the other two zinc

Figure 3.11. Amino acid sequence of DNA-binding domain of glucocorticoid receptor used for crystallization with the DNA response element. The numbering corresponds to that of the full-length native receptor. The boxed residues are α-helical. Residues that make contact at the dimer interface are indicated by the solid dots (477, 479, 481, 483, and 491). Residues making phosphate contacts (451, 452, 475, 489, 490, and 496) are indicated by solid rectangles. The arrows indicate base contacts. A disordered section at the carboxyl terminus is indicated by the dashed lines. Three amino acid residues that direct the discrimination between glucocorticoid and estrogen response elements are indicated by white lettering in the solid boxes (458, 459, and 462); those that discriminate between estrogen and thyroid response elements are circled (478–481). The lower-case letters indicate artifacts of the expression vector construct. (Reproduced, by permission, from Luisi et al., 1991)

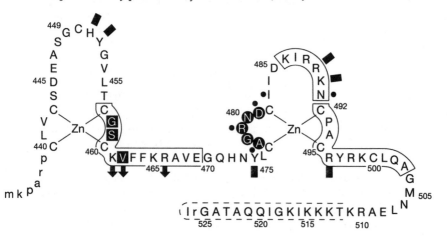

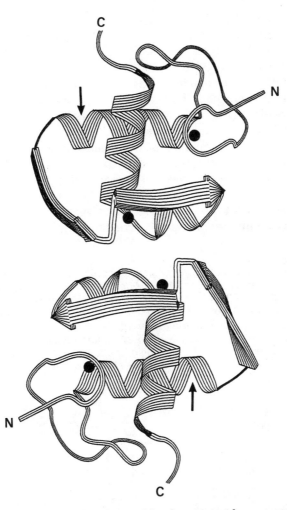

Figure 3.12. The DNA-binding dimer of the glucocorticoid receptor viewed toward the DNA. The black circles mark the zinc ions. The arrows point to the α-helices that nestle in two successive grooves of the DNA double helix. (Courtesy of Dr. Paul Sigler)

fingers bind to two successive major grooves of the DNA double helix, with their long axes normal to its axis. In the complex with the natural response element where the palindromic base pairs are separated correctly by three base pairs, the binding of the two α-helices is symmetrical, showing that the separation between them is made to measure for binding to two successive major grooves of the DNA. Unfortunately, the resolution is insufficient to see the detailed interactions between the DNA and the protein.

In the second complex, whose structure is well resolved, only one of the α-helices is properly positioned in a major groove, but the other α-helix cannot reach the base of the next major groove (Fig. 3.13). The authors regard the binding of the first α-helix as representative of the binding of the two α-helices in the symmetrical structure, and they describe its contacts with the DNA in detail (Fig. 3.14). The contacts consist mainly of hydrogen bonds from arginines, lysines, and tyrosines to phosphates and bases of the DNA. Six of the contacts are with residues that are part of the N-terminal α-helix and zinc finger; three are with tyrosine 474 and arginines 489 and 496 that are part of the C-terminal zinc finger. The hydrogen bond formed by arginine 489 must be crucial, because this residue is

Figure 3.13. Stereo view of the glucocorticoid receptor protein–DNA complex. The dark circles mark the zinc ions. The domain at the top is the one that is correctly positioned. Protein sidechains that form hydrogen bonds with the DNA are indicated. The α-helix in the major groove is viewed end-on. The residues in the top domain are numbered by the last two digits of the sequence (440-525); those in the bottom domain by subtracting 300 from that sequence. Only the binding of the top domain is likely to be representative of that of the two domains in the natural complex.

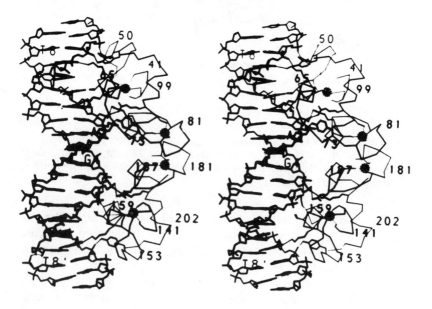

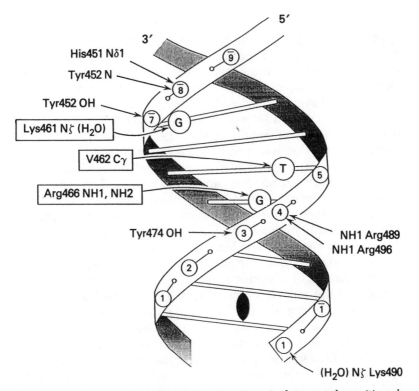

Figure 3.14. Diagram of protein–DNA interactions in the correctly positioned domain of the glucocorticoid receptor protein-DNA complex. Numbered circles indicate phosphates. (Reproduced, by permission, from Luisi et al., 1991)

conserved in many receptors and its replacement by glutamine in the human vitamin D receptor was found to be the cause of one family's hereditary vitamin D–resistant rickets. A van der Waals contact that valine 462 makes with thymine 5 may be crucial for discrimination between the glucocorticoid and estrogen receptors, where this thymine is replaced by a cytosine and the valine by an alanine.

From Fruit Flies to Leukemia: Homeoboxes and Homeodomains

In 1965 Walter Gehring, then a graduate student in Zürich, discovered a mutant of the fruit fly *Drosophila* that had grown legs instead of antennae on its head. It was one of several mutants in which one part of the fly's body had grown on another where it did not belong. These mutants were called *homeotic* (from *homoios*, Greek for alike) because they affected genes that act alike. They all appeared to orchestrate the expression of groups of other genes responsible for the ordered development of the fly. Gehring isolated and cloned such a gene, called *antennapedia*, that determines the development of the fly's head and anterior thoracic segments. When cDNA derived from this gene was used as a probe to hybridize with pooled *Drosophila* DNA, many other genes containing similar base sequences were extracted. Since they were part of homeotic genes, they were named *homeoboxes* (Gehring, 1985, 1987).

The name was appropriate, because their nucleotide base sequences exhibited strong homology. One section of 180 among the antennapedia gene's 360-odd base pairs coded for a domain of almost invariant amino acid sequence. That homeodomain has since been found in the genomes of most eukaryotes, from yeast to man, and its sequence has been faithfully conserved throughout evolution. For example, one homeodomain in *Drosophila* has 59 of its 60 amino acid residues in positions identical to those in a domain of the toad Xenopus (Fig. 3.15). All 16 homeodomain sequences of *Drosophila* are flanked by polypeptide sequences that exhibit lesser, but still strong, homology (Gehring, 1987).

In yeast, certain genes were found to code for proteins that regulate differentiation into either of two mating types or into spores. These proteins contain a sequence of amino acid residues that is partially homologous to the homeodomains of *Drosophila* and that includes similar clusters of arginines and lysines. The proteins bind to yeast DNA 5′ to the genes whose expression they regulate. This observation suggested that all homeoboxes control gene expression by binding to specific DNA sequences, a function that would make them the analogues of the prokaryotic repressors and activators to some of which they are, in fact, weakly homologous (Table 3.1). That hint was corroborated by the localization of

Figure 3.15. (opposite) Comparison of the amino acid sequences encoded by the homeobox genes of Drosophila, honeybee, and mouse. Labels are: en—engrailed; inv—invected of Drosophila; E60 and E30—homologous sequences of the honeybee; En-1 and En-2—homologous sequences of the mouse. The most frequently found amino acids among the six sequences are indicated in lighter shaded regions for the homeodomain and in darker shaded regions for the flanking sequences that show homology. The amino terminal sequences are not shown; insertions and deletions are indicated in parentheses. The arrowheads indicate the location of an intron in the two Drosophila genes. The stars refer to the carboxyl termini of the proteins. (Reproduced, by permission, from Gehring, 1987)

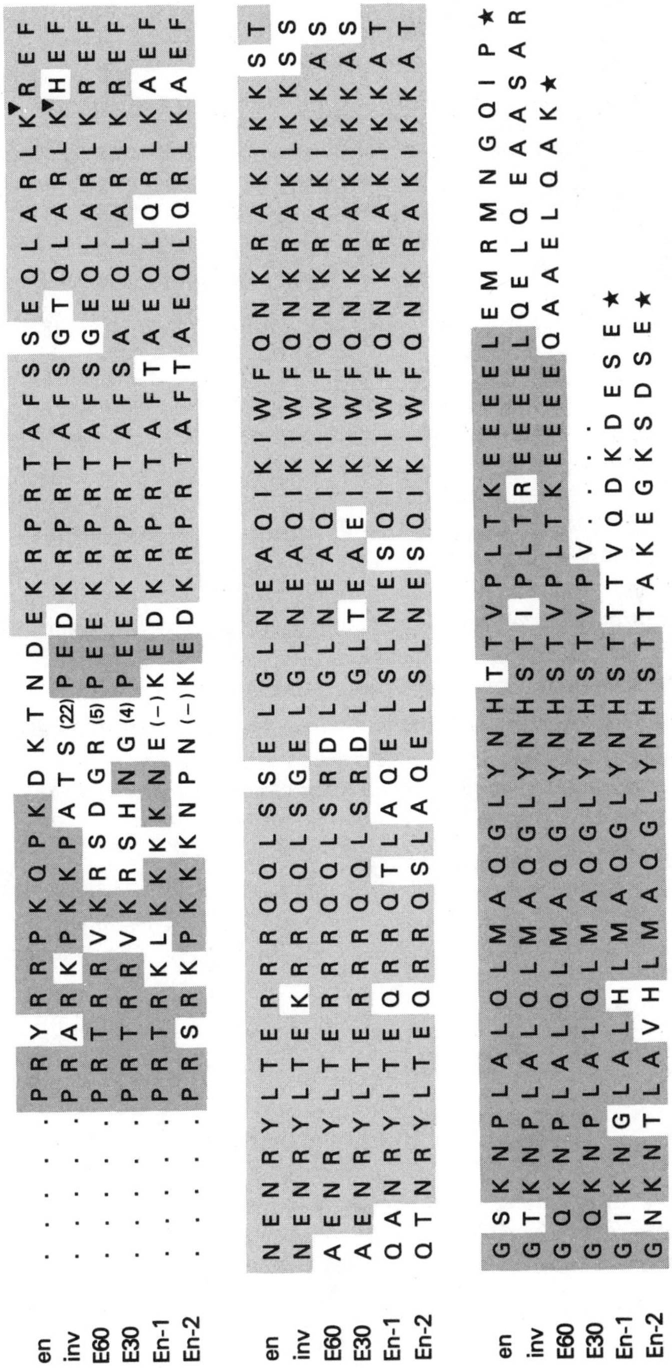

```
en    . . . . . P R Y R R P K Q P K D K T N D E K R P R T A F S S E Q L A R L K R E F
inv   . . . . . P R A K P K K P A T S (22) P E D K R P R T A F S G T Q L A R L K H E F
E60   . . . . . P R T R R V K R S D G R (5) P E E K R P R T A F S G E Q L A R L K R E F
E30   . . . . . P R T R R V K R S H N G (4) P E E K R P R T A F S A E Q L A R L K R E F
En-1  . . . . . P R T R K L K K K K N E (-) K E D K R P R T A F T A E Q L Q R L K A E F
En-2  . . . . . P R S R K P K K N P N (-) K E D K R P R T A F T A E Q L Q R L K A E F

en    N E N R Y L T E R R R Q Q L S S E L G L N E A Q I K I W F Q N K R A K I K K S T
inv   N E N R Y L T E K R R Q Q L S G E L G L N E A Q I K I W F Q N K R A K L K K S S
E60   A E N R Y L T E R R R Q Q L S R D L G L N E A Q I K I W F Q N K R A K I K K A S
E30   A E N R Y L T E R R R Q Q L S R D L G L T E A E   I K I W F Q N K R A K I K K A T
En-1  Q A N R Y I T E Q R R Q T L A Q E L S L N E S Q I K I W F Q N K R A K I K K A T
En-2  Q T N R Y L T E Q R R Q S L A Q E L S L N E S Q I K I W F Q N K R A K I K K A T

en    G S K N P L A L Q L M A Q G L Y N H T T V P L T K E E E E L E M R M N G Q I P ★
inv   G T K N P L A L Q L M A Q G L Y N H S T I - P L T R E E E L Q E L Q E A A S A R
E60   G Q K N P L A L Q L M A A G L Y N H S T V P L T K E E E E Q A A E L Q A K ★
E30   G Q K N P L A L Q L M A A G L Y N H S T V P V . . .
En-1  G I K N G L A L H L M A Q G L Y N H S T T V Q D K D E S E ★
En-2  G N K N T L A V H L M A Q G L Y N H S T T A K E G K S D S E ★
```

Figure 3.15.

Table 3.1. Amino acid sequences of DNA-binding motifs of λ-phage repressor, Drosophila homeodomain, and human oncogene

	HELIX 2	TURN		HELIX 3	
	36	40 41 42	47	50 51 52
λ-repressor	-Val-X-X-X-	Met-Gly-Met-	X-X-X-X-Val-	X-X-	Leu-Phe-Asn-
Homeodomain	-Leu-X-X-X-	Leu-Gly-Leu-	X-X-X-X-Ile-	X-X-	Trp-Phe-Gln-
Prl oncogene	-Leu-X-X-X-	Cys-Gly-Ile-	X-X-X-X-Val-	X-X-	Trp-Phe-Gly-
	34	38 39 40	45	48 49 50

antihomeodomain antibodies in cell nuclei and, *in vitro*, by studies that proved their binding to specific base sequences within the clusters of *Drosophila* genes which they control. It has since become clear that they regulate transcription of genes in many higher organisms: they represent one of several types of eukaryotic transcription factors (Gehring, 1987).

The structures of the homeodomain of the antennapedia gene of *Drosophila* and of its complex with its DNA receptor have been solved

Figure 3.16. Structure of antennapedia homeodomain of Drosophila. *Stereoview of one of 19 related structures obtained from proton–proton distances determined by NMR, followed by distance geometry calculations and restrained energy refinement (restrained to keep interatomic distances and bond angles within correct limits). Only the backbone and the sidechains of conserved residues are shown. N—asparagine; H—histidine; F—phenylalanine; R—arginine. (Reproduced, by permission, from Qian et al., 1989)*

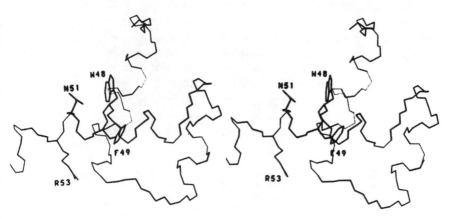

by nuclear magnetic resonance spectroscopy (Qian et al., 1989; Billeter et al., 1990; Otting et al., 1990) and the structure of the homeodomain of the *engrailed* gene of *Drosophila* complexed with its DNA receptor, by X-ray crystallography (Kissinger et al., 1990). The two structures are very similar. Figure 3.16 shows the overall fold of the polypeptide chain first seen as a result of the nuclear magnetic resonance work. It is made up of three helices linked by irregular loops. The second and third helices and the loop connecting them are reminiscent of the helix-turn-helix motif found in prokaryotic DNA-binding proteins such as the λ-phage repressor of *E. coli* (Ptashne, 1986; Sauer et al., 1982).

Figures 3.17 and 3.18 show the X-ray structure of the DNA-binding domain of the λ-repressor bound to its operator (Jordan and Pabo, 1988). Figure 3.19 shows similar views of the *engrailed*

Figure 3.17. Interaction of phage λ-repressor dimer with DNA operator. Each monomer contains the helix-turn-helix motif. Helix 3 snuggles into the major groove of the DNA double helix. (Reproduced, by permission, from Jordan and Pabo, 1988)

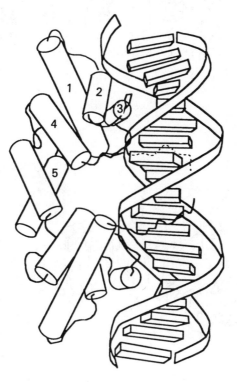

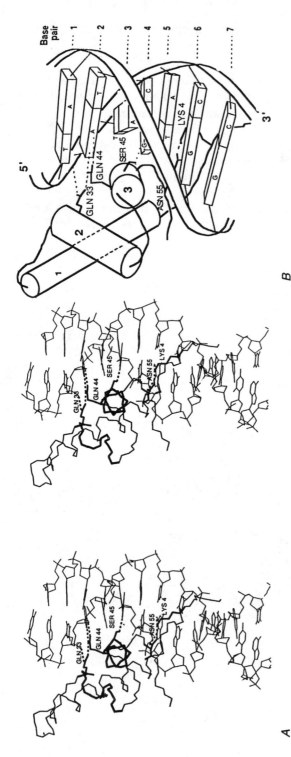

A

B

Figure 3.18. Interaction of one of the λ-repressor monomers of E. coli with the λ-operator DNA. (A) Stereodiagram showing the sidechains that interact with base pairs in the major groove. The backbone of the protein is shown for residues 1 to 58; the NH₂-terminal arm begins in the lower-right corner of this figure. Critical sidechains and the backbone of helices 2 and 3 are emphasized with bold lines. (B) Sketch, in the same orientation as A, showing hydrogen bonds with the DNA. In the upper half of the figure, the major groove is readily visible for base pairs 1 to 3. In the lower half, the minor groove is closest to the viewer.

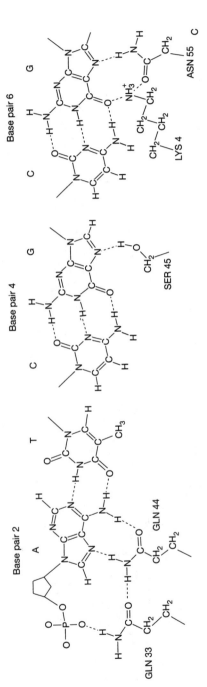

(C) Sketches summarizing how sidechains form hydrogen bonds with base pairs 2, 4, and 6. The solid lines show the connection to the sugar phosphate backbone. Since Gln 33 actually hydrogen bonds to the backbone, the sketch of base pair 2 shows the phosphate on the 5' side of the adenine. (Reproduced, by permission, from Jordan and Pabo, 1988)

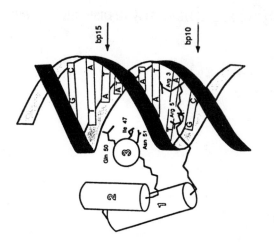

B

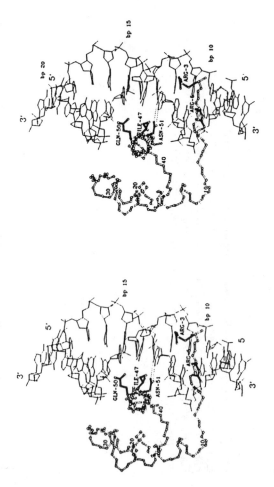

A

Figure 3.19. Interaction of the engrailed homeodomain of Drosophila with its homeobox
DNA receptor. Key contacts with the base pairs are at the TAAT subsite. (A) Stereodiagram
showing a view along helix 3. Backbone atoms are shown for residues 3–59 of the
homeodomain. Sidechains are shown for residues that contact the base pairs: Arg-3,
Arg-5, Ile-47, Gln-50, and Asn-51. The segment of DNA includes base pair 8 (at the
bottom) through base pair 20 (at the top) and thus includes the critical TAAT subsite.
Base pairs 10, 15, and 20 are labeled, along with the 3′ and 5′ termini of each DNA
strand. The N-terminal arm makes contacts with the minor groove near the 5′ end of
the TAAT subsite. Helix 3 makes contacts with the major groove near the 3′ end of the
TAAT subsite. (B) Sketch summarizing the critical contacts. (Reproduced, by permission,
from Kissinger et al., 1990)

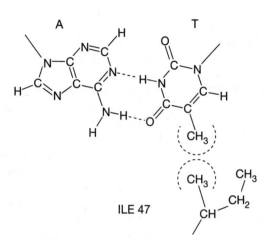

ILE 47

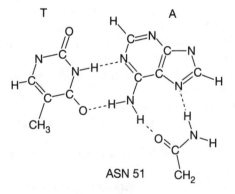

ASN 51

Figure 3.19. (C) Sketch showing the base contacts made by Ile-47 and Asn 51. (Reproduced, by permission, from Kissinger et al., 1990)

homeodomain bound to its homeobox (Kissinger et al., 1990). The similarities are striking, except that the λ-repressor, in common with other prokaryotic repressors, binds as a dimer, while homeodomains bind as monomers. The helices of both proteins are held together by a core of hydrophobic sidechains in equivalent position. The turn of the loop connecting helices 2 and 3 is facilitated by the same glycines (41 and 39). The sidechains of tryptophan 48 and phenylalanine 49, which are invariant in the homeodomains of higher eukaryotes, protrude from the hydrophobic side of helix 3 (Fig. 3.16). Their equivalents in the phage repressor are leucine 50 and phenylalanine 51.

In both proteins, the other, mainly hydrophilic, side of helix 3 binds to the major groove of a double-helical DNA oligonucleotide. Along this sequence an invariant asparagine (51) in the homeodomain forms two hydrogen bonds with an adenine. Serine 45 is its equivalent in the phage repressor. In the homeodomain, a glutamine (50) is placed so that a small change in the conformation of the DNA would allow it to form two hydrogen bonds with another adenine, as does the equivalent glutamine 44 in the phage repressor. The sidechains of isoleucine 47 of the homeodomain and alanine 49 of the repressor make equivalent contacts with methyl groups of thymines. The N-terminal arm of the homeodomain is wrapped around the double helix, allowing arginines 3 and 5 to form hydrogen bonds with thymines in the minor groove (Figs. 3.18 and 3.19). The N-terminal arm of the phage repressor also wraps around the double helix, but its basic sidechains bind to bases in the major groove. Despite these similarities, helices 3 of the two proteins occupy different positions in the major grooves of their DNA receptors: that of the phage repressor is buried more deeply and shifted in the direction of its C-terminus relative to that of the homeodomain. Considering the immense distance on the evolutionary scale that separates the two structures, their similarities are more remarkable than their differences.

Homeobox domains recognize their DNA receptors with high specificity: replacement of glutamine 50 by lysine changes the affinity of *Drosophila* homeodomains from one set of homeoboxes to another (Treisman et al., 1989).

Until 1990 these discoveries appeared to be of mainly biological interest. In that year two groups discovered independently that a human homeobox plays a part in the pathogenesis of an acute lymphoid leukemia (Nourse et al., 1990; Camps et al., 1990). Cytogenetic studies had revealed chromosomal translocations in the malignant lymphoid cells of nearly half such leukemias. They were of different kinds, but they all resulted in the conversion to oncogenes of proto-oncogenes concerned with growth and differentiation. (For definitions of oncogenes and proto-oncogenes, see Chapter 8.)

The translocations are due to the crossing over and rupture of the DNA in two adjacent chromosomes and the rejoining of the free strands. t(1;19) signifies such a translocation that has led to the reciprocal exchange of DNA between chromosomes 1 and 19. Translocations of this kind have been found in several human leukemias. In the present case, chromosome 19 had broken within a gene called E2A. This gene codes for two factors that enhance transcription of immunoglobulin genes. They are proteins E12 and E47.

The translocated gene contains the first 1519 bases of the E2A gene with an insertion of three extra bases at positions 1236–1238. After base 1519 the sequence switches to 1028 nucleotides derived from band 23 on chromosome 1. The fusion sequence of 2550 nucleotides coded for a protein of 85 kilodaltons, as compared to a 68-kilodaltons protein coded for by the E2A gene alone. As a result of the translocation, the E2A protein had lost its helix-turn-helix and dimer-forming motifs and had acquired instead from chromosome 1 part of a new protein called *Prl* (for pre–B-cell leukemia) that included a sequence of 64 residues markedly homologous to homeodomains of *Drosophila* (Table 3.1). In other words, the normal DNA-binding domain of the enhancing factor E2A was replaced by one with the wrong specificity, one that was likely to activate transcription of some wrong gene or genes.

Only a little more than 20 of the Prl homeodomain's residues are identical to those that are also identical among all the other known homeobox sequences, and optimal alignment needs three extra residues to be inserted in the Prl domain. However, the characteristic sequence tryptophan–phenylalanine–X–asparagine–X–arginine is present. Five other key residues, including two important arginines, also appear in identical positions and a sequence of four arginines and a lysine near the N-terminus is well placed to bind to phosphates in the minor groove of a double helix (Table 3.1 and Fig. 3.20). Prl may therefore represent the first member of a new family of sequence-specific DNA-binding proteins that regulate development. The structure of the Prl protein and identity of the gene or genes activated by it have yet to be determined. Once known, they may open new avenues to the treatment of leukemias.

The t(1;19) translocation described here is associated with a specific subclass of acute lymphoblastic B-cell leukemia in children. Translocation of an antennapedia-like homeobox adjacent to the chromosomal breakpoint has also been found in a lymphomatous T-cell acute leukemia in a five-year-old child (Kennedy et al., 1991).

In 1991 a sequence homologous to the antennapedia homeodomain of *Drosophila* was found in a nuclear protein extracted from the bone marrow of a baboon fetus. The protein binds to an A-T-rich region of the promoter of the γ-globin genes and to a nucleotide base sequence 3′ to the ^Aγ-gene that acts as an enhancer of globin messenger RNA transcription. It also binds to a locus control region

	HELIX	TURN	HELIX	
Prl	DARRKRRNFNKQATEILNEYFYSHLSNPYPSEEAKEELAKKCGITVSQVSNWFGNKRIRYKKNI			human
		: :	: *** * . :	
Oct2	RR-K--TSIETNVRFA-EKS-	-A-QK-TS-EILLI-EQLHMEKEV-RV--C-R-QKE-RIN		human
Pit1	RK-KR-TTISIA-KDA-ERH-	GEHSK--SQEIMRM-EELNLEKEV-RV--C-R-Q-E-RVK		rat
c1	XRK-G-QTYTRYQ-LE-EKE-	HY-R-LTRRRRI-I-HALCL-ER-IKI--Q-R-MKW--EN		human
Hu1	-GK-A-TAYTRYQ-LE-EKE-	HF-R-LTRRRRI-I-HALCLSER-IKI--Q-R-MKW--EN		human
huhox2.5	XS-K--CYPT-YQ-LE-EKE-	-F-M-LTRDRRH-V-RLLNLSER--KI--Q-R-MKM--MN		human
huhox2.7	XSK-A-TAYTSAQLVE-EKE-	HF-R-FVRPRRV-M-NLLNLSER-IKI--Q-R-MQ---DQ		human
huhox2.4	XR--G-QTYSRYQ-LE-EKE-	-F-P-LTRKRRI-VSHAL-L-ER--KI--Q-R-MKW--EN		human
k8	E---L-TAYTNTQLLE-EKE-	HF-K-LCRPRRV-I-ALLDL-ER--KV--Q-R-MKH-RQT		human
en	-EK-P-TA-SSEQLAR-KRE-	NE-R-LT-RRRQQ-SSEL-LNEA-IKI---Q---AKI--ST		fly
ant	ERK-G-QTYTRYQ-LE-EKE-	HF-R-LTRRRRI-I-HALCLTER-IKI--Q-R-MKW--EN		fly
MATa1	KSPKGKSSISP--RAF-EQV-	RRKQSLNSKE---V-------PL--RV--I---M-S-		yeast
MATa2	TKPYRGHR-T-ENVR--ESW-AKNIE---LDTKGL-N-M-NTSLSRI-IK--VS-R-RKE-TIT			yeast
Consensus	R Y Q L F Y R A L L Q KIWFQNRR K K			

Figure 3.20. Similarity of amino acid sequence of the Prl fusion protein predicted from homology of the nucleotide base sequence of its cDNA with those of other homeodomains. Dashes indicate amino acid identities with Prl. The consensus sequence in the last line gives residues conserved in homeodomains throughout the higher eukaryotes. The four invariant amino acids in all non-yeast homeodomains are denoted by asterisks. Identical matches or conservative changes between Prl and eight other highly conserved homeodomain residues are denoted by colons and periods, respectively. Amino acids that form the proposed helix-turn-helix motif are indicated at the top. A gap of three amino acids has been introduced in all but the yeast MATa2 sequence to maximize the sequence similarities with Prl. (Reproduced, by permission, from Nourse et al., 1990)

20 kilobases 5′ to the embryonic ε-globin gene that confers high levels of expression to that gene (Lavelle et al., 1991). This is just one more example of the many roles that homeodomains play in the control of gene expression in eukaryotes. It is also becoming clear that they function as oncogenes in many kinds of tumors where their role as transcriptional regulators has been perverted.

The Leucine Zipper and SPXX Motifs in Cell Regulation and Oncogenesis

Two other structural motifs in proteins contribute to gene regulation and carcinogenesis: the leucine zipper and the sequence motif serine- or threonine-proline-X-X-. Landschulz and others discovered that three transforming proteins and two transcriptions factors (i.e., proteins that regulate transcription of DNA into messenger RNA) share a motif that consists of a repeat of several leucines at intervals of seven residues (Landschulz et al., 1988). In an α-helix that contains 3.6 residues per turn, leucine sidechains seven residues apart would repeat along one face in an almost straight row at intervals of 10.8Å (Fig. 3.21). McKnight and his colleagues proposed that such a helix on the surface of one protein molecule would associate with another helix in a second molecule to form a dimer in which the leucine sidechains of adjacent helices interdigitate in a way Crick suggested in 1952. Crick showed that sidechains protruding from one helix would not pack into the spaces between the sidechains of its neighbor unless the two α-helices were inclined to each other at 20° and wound around each other to form a coiled coil (Crick, 1952).

O'Shea and others (1991) have confirmed Crick's model in the crystal structure of a peptide 33 amino acids long, cleaved from the C-terminus of the yeast transcriptional activator GCN4. This α-helical peptide contains four leucines spaced seven residues apart (see Fig. 3.22A). Two such helices join together in parallel at an inclination of 18° between their helical axes to form one quarter of a coiled coil. The pair looks like a twisted ladder with sides made of the peptide backbones and rungs made of the hydrophobic sidechains between them (Fig. 3.22B). Figure 3.23 on page 109 shows how the mutal inclination of the two helices by 20° allows any one sidechain of either helix to pack into the space between four sidechains of its partner helix. The two helices are also linked by salt bridges between three pairs of glutamates and lysines and by a hydrogen bond between two asparagines—perhaps because

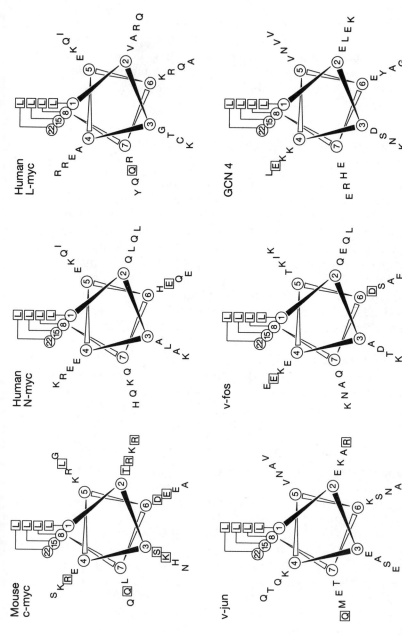

Figure 3.21. Repeat of leucines at intervals of seven residues along hypothetical α-helices in five oncogene and proto-oncogene products and in the yeast transcription activator GCN4. The residues are marked in single letter code. (See Figure A2.1 on page 278.) (Reproduced, by permission, from Landschulz et al., 1988)

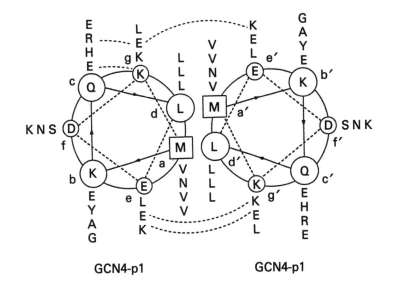

GCN4-p1   GCN4-p1

GCN4-p1	Ac-R	MKQ	LEDKVEE	LLSKNYN	LENEVAR	LKKLVGE	R
heptad position		a b c	d e f g a b c	d e f g a b c	d e f g a b c	d e f g a b c	d

Figure 3.22. (A) Amino acid sequence of residues 249–281 of the yeast transcription activator protein GCN4 and, above, its plot on a helical wheel. Going clockwise, the first methionine is boxed and the residues in the first two turns are circled. Successive residues n × 7 places below those in the first two turns are listed above, below, or beside those in the first two turns. Heptad positions are labeled a–g. Leucines at d interact with residues d′ and e′. Alternate layers contain residues at a and a′, and d and d′. Dashed lines mark salt bridges between lysines and glutamates.

the hydrophobic bonds alone are not strong enough to link two large proteins together.

Exactly six residues N-terminal to the first of the leucines in leucine zipper proteins lies a region of 16 residues rich in arginines and lysines that is essential for binding them to DNA (Kouzarides and Ziff, 1988). Vinson, Sigler, and McKnight (1989) have proposed that the zipper and its adjoining basic region are shaped like a T: its vertical stroke represents the coiled-coil of the zipper and its horizontal bar the basic regions on either side that wrap around the major groove of DNA in opposite directions. The distribution of arginines and lysines varies in different leucine zipper proteins, but several of the sequences suggest that they form part

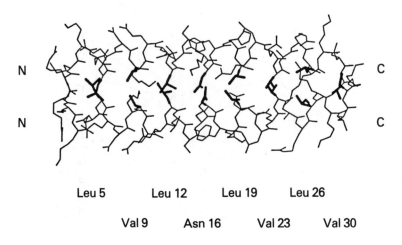

N   C

N   C

Leu 5	Leu 12	Leu 19	Leu 26
Val 9	Asn 16	Val 23	Val 30

Figure 3.22. (B) Side view of the double helix of protein GCN4, showing the ladder of leucine, valine, and asparagine sidechains between the two helices in bold.

of β-strands rather than α-helices as Vinson and others suggested. For example, one of the zipper proteins contains the sequence Arg-Asp-Lys-Ala-Lys-Glu-Arg. In an α-helix the distance between two positively charged sidechains four residues apart would be only 4Å, so that they would repel each other; besides, only half of them could face the DNA. In a β-strand, on the other hand, they would lie 7Å apart all on one side facing the DNA, while the sidechains between them would face away from the DNA. Other leucine zipper proteins contain the sequences Arg-Asn-Arg-Arg-Arg or Arg-Arg-Lys-Lys-Lys. Again, electrostatic repulsion would be minimized, and attraction to DNA maximized if they formed part of β-strands.

Suzuki discovered that the sequence serine- or threonine-proline-X-X- occurs at a higher frequency in DNA-binding proteins than in other proteins, and that such SPXX or TPXX sequences often flank helix-turn-helix or zinc finger domains (Suzuki, 1989). Steroid hormone receptors contain between 4 and 21 such motifs, DNA polymerases between 26 and 52, and histone H1 many where X are basic residues. Suzuki suggested that the motif may form type I β-turns in which the NH of the serine or threonine points outward while the serine OH forms an internal hydrogen bond that stabilizes the β-turn (Fig. 3.24). (See also Figure A2.7 on page 289) Suzuki

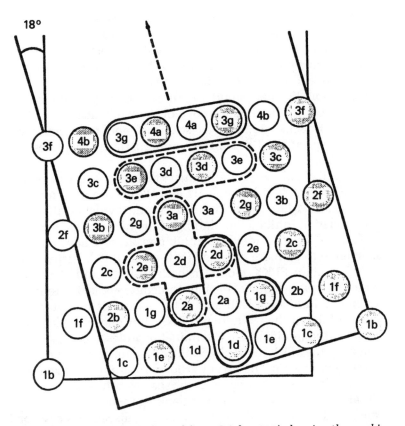

Figure 3.23. Helical net (adapted from Crick, 1953) showing the packing of knobs from one helix into holes of its partner in the protein GCN4. The net was obtained by wrapping a piece of paper around each helix, marking the α-carbon positions with circles, and then superimposing the two unwrapped sheets with the correct contacts. Open circles represent the C_a's of one helix, shaded circles represent those of its partner. Residue 2a fits into the hole between 2a, 2d, 1g, and 1d from the partner helix, and so on. The numbers represent helical turns. The letters represent the heptad repeats shown in Figure 3.22A. (Reproduced, by permission, from O'Shea et al., 1991)

then proposed that a row of SPXXs would line up along and bind to the minor groove of double helical DNA in the same way as the drug distamycin shown in Figure 4.7 on page 135, such that the NH groups of the serines donate hydrogen bonds to adjacent bases.

Churchill and Suzuki (1989) tested this proposal experimentally. They studied the protection of DNA from attack by hydroxyl radicals when it was bound either to a 49-residue peptide from sea urchin histone H1 that contained six repeats of a sequence -serine-proline-(lysine or arginine)$_2$-, or to a synthetic peptide with the sequence (serine-proline-arginine-lysine-)$_2$, or to the drug Hoechst 33258, which binds in the same manner as distamycin. The three different ligands bound to similar adenine-thymine–rich base sequences in the minor groove of double-helical DNA. The binding of Hoechst 33258 was strongest and most narrowly localized to specific A-T pairs. The binding of the two peptides was weaker and more diffuse.

Suzuki and others discovered a kinase that specifically phosphorylates SPKK motifs in histones H1 and H2B of sea urchin sperm (Suzuki et al., 1990). Suzuki (1991) found evidence that this phosphorylation inhibits the binding of the histones to DNA, but when Hill and others studied that inhibition closely, they found the phenomenon to be more complex (Hill et al., 1991). Histones H1 and H2B contain SPKK motifs in both their N- and C-terminal peptides. Phosphorylation of the four tandemly repeated N-terminal SPKK motifs in histone H2B and of the six motifs in histone H1 strongly inhibited their binding to DNA, but phosphorylation of three dispersed motifs in the C-terminal peptides had little effect. In summary, the SPKK motif clearly does bind to, or at least covers, AT-rich sequences in the minor groove of double-helical DNA, but the binding strength varies with the spacing between successive motifs and probably also with their flanking sequences.

The functions of leucine zippers and their adjacent basic regions and SPXX motifs have emerged most clearly from the study of the products of two proto-oncogenes called *c-Jun* and *c-Fos* (Boyle et al., 1991; Ransone and Verma, 1990). The normal cellular proteins c-Fos and c-Jun can induce tumors *in vivo* and transform fibroblasts *in vitro*. c-Jun contains 334 and c-Fos contains 380 amino acid residues. Each contains one leucine zipper preceded by many basic residues (Kouzarides and Ziff, 1989).

In vitro c-Fos and c-Jun associate to form heterodimers, which was suspected to be due to the mutual attraction of their leucine zippers. This was confirmed when synthetic peptides with the sequences of the leucine zippers of c-Fos and c-Jun formed heterodimers with a dissociation constant that was about a thousand times smaller than that of the homodimers formed by either peptide alone (O'Shea et al., 1989). The experiment showed that pairing by leucine zippers is specific; this might be due to juxtaposition of complementary polar groups.

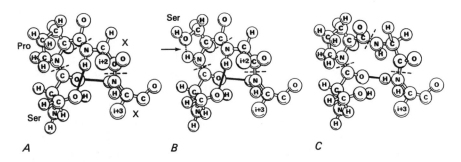

Figure 3.24. Proposed structures of serine-proline-X-X sequence motif. They represent variations of β-turns. In (B) the proline in position 2 is replaced by another serine whose OH accepts a hydrogen bond (arrow) from its own mainchain NH, thus mimicking a proline ring. The thick lines are other hydrogen bonds and the broken lines borders between successive residues. The arrow in (A) points to the serine NH that is thought to form hydrogen bonds with adenines in the minor groove of double helical DNA. Sidechains for the third and fourth residues have been omitted except for the C_β atoms. (Reproduced, by permission, from Suzuki, 1989)

The Jun and Fos proteins contain neither zinc fingers nor helix-turn-helix motifs. Instead, they appear to owe their affinity to DNA to the concentrations of basic residues that precede their leucine zippers and to SPXX motifs. DNA-binding is regulated by a complex mechanism: by their association into heterodimers via their leucine zippers, by phosphorylation and dephosphorylation of the serines in their SPXX motifs, and by oxidation and reduction of their cysteines. The experiments described next show that formation of heterodimers and dephosphorylation strengthen, while dissociation into monomers and phosphorylation weaken their binding to their DNA receptors.

The phorbol ester 12-0-tetradecanoylphorbol-13-acetate or TPA promotes growth of tumors in association with carcinogens. The mechanism of promotion has been tracked down to the binding of a large polymeric protein to a heptanucleotide with the sequence TGA(C or G)TCA known as the TPA response element or TRE. Binding of the protein to TRE activates transcription of messenger RNA for a complex of other proteins involved in oncogenesis. Jun and Fos are part of that polymeric protein complex, yet their binding to TRE is alone sufficient to activate transcription. Dimers of Jun alone can do so; heterodimers of Jun and Fos do so more efficiently, because they are far more stable, but whether they do so or not depends on their state of phosphorylation (Ransone and Verma, 1990).

c-Fos contains the sequence -arginine 259-lysine 260-glycine 261-serine 262–264-. When its serines are phosphorylated, transcription of the Fos-promotor gene adjacent to the TPA response element is repressed. It is similarly repressed when the serines are replaced by aspartates and glutamates. Dephosphorylation of the serines stimulates transcription (Ofir et al., 1990).

The latent form of c-Jun is strongly phosphorylated and binds to the TPA response element TRE only weakly. TPA activates its binding to TRE by inducing the dephosphorylation of it serines. The binding can be repressed by reaction of c-Jun with glycogen-synthase kinase that phosphorylates the serines and threonines at its proline-rich sites. Phosphorylation is largely inhibited by the replacement of a single serine (243) by phenylalanine. The mutant c-Jun carrying that replacement bound to TRE 10 times more strongly than wild type c-Jun. It may therefore be significant that in the v-Jun protein of the avian sarcoma virus, a phenylalanine replaces the serine at the central phosphorylation site of the normal avian c-Jun protein (Boyle et al., 1991).

Phosphorylation of the serines in the SPXX motifs breaks the hydrogen bond which stabilizes the β-turns and also generates electrostatic repulsion of DNA. The combined effects inhibit the binding of Fos and Jun to their DNA receptors. Similarly, phosphorylation of serines and threonines is likely to detach histone H1, DNA polymerase II, and other proteins from DNA, while dephosphorylation would promote their binding.

Abate and others have discovered yet another mechanism that regulates the binding of the Jun-Fos heterodimer to its DNA receptor (Abate et al., 1990) The amino acid sequences of each of the two proteins contains one cysteine that precedes and another that follows the leucine zipper. Reaction of the cysteines with the SH-reagent N-ethyl-maleiimide, oxidation of the SH-groups, or replacement of the cysteines by methionines inhibited DNA binding. Incubation of the protein with an oligonucleotide containing the TRE sequence inhibited the reaction of the cysteines with N-ethylmaleiimide, which shows that the DNA covers the cysteines. Replacement by serine of only the cysteine that precedes the leucine zipper enhanced the binding to DNA, even in the presence of strong oxidizing agents. Those cysteines are part of a conserved trio lysine-cysteine-arginine, suggestive of being a DNA-binding motif. In the presence of a nuclear liver extract, DNA binding of Jun-Fos was enhanced by the enzyme thioredoxin, especially on addition of thioredoxin reductase and

NADPH. The enhancement was best at 37°C, weaker at 24°C, and absent at 4°C. From these observations, the authors infer the presence of a nuclear enzyme that catalyzes the reduction of the sulphydryl groups of the Fos and Jun proteins and thereby enhances their binding to TRE.

The possible nature of that enzyme has emerged from another family of oncogenes that are translocated in children suffering from acute lymphoid leukemia. Several authors had found mammalian homeobox domains to be preceded by a cysteine-rich domain, called LIM, that contains the duplicated motifs histidine-X_2-cysteine-X_2-cysteine-X_2-cysteine, followed by a single cysteine further along the sequence (Karlsson et al., 1990; Freyd et al., 1990; McGuire et al., 1989). Boehm and others found such a LIM domain coded for by a gene that had undergone a chromosomal translocation in acute lymphoid leukemia (Boehm et al., 1991). They recognized the sequence cysteine-X_2-cysteine-X_2-cysteine-X_N-cysteine (where N is a variable number of residues) as one that occurs in the structural zinc binding site of alcohol dehydrogenase and in several ferredoxins. These are redox enzymes that contain clusters of iron and inorganic sulfur, Fe_4S_4, arranged at the corners of a distorted cube. Three closely linked cysteines of the protein bind to iron atoms at three corners of a cube and the more distant cysteine binds to the fourth. In LIM proteins the histidine that precedes the three cysteines may be the fourth metal ligand.

The putative ferredoxin domain found to be part of the homeobox and the LIM proteins may be responsible for the nuclear activity which catalyzes the reduction of the cysteines in Jun and Fos. In that case ferredoxin may be yet another protein regulating gene transcription.

Hormones and many other chemical signals induce cellular responses by activating kinases on the cell surface. The kinases in turn activate other kinases, which transmit signals to proteins in the cell nucleus that regulate transcription. For example, a protein kinase called $p34^{cdc2}$ initiates mitosis and meiosis (M-phase) by phosphorylating nuclear proteins at SPXX sites (Moreno and Nurse, 1990; Broek et al., 1991). Activation or repression often requires the linkage and cooperation of more than one nuclear protein. Jun and Fos, together with the leucine zipper and SPXX motifs, provide the first molecular model of such cooperation and interaction with specific genetic loci in health and disease. Therein lies the fundamental importance of the results reported in this section.

Conclusions

DNA and proteins have often learned to recognize each other by evolving palindromic nucleotide base sequences and surfaces of dimeric proteins that are complementary to them. When DNA and protein combine, the same axis of twofold symmetry that brings one half of the palindrome into coincidence with the other by a rotation of 180° also relates the positions of the two protein monomers that have coalesced into a dimer.

Such an arrangement has several advantages. A globular protein molecule made up of no more than a hundred amino acid residues would adhere to DNA over too small a surface area to bind with an energy large enough to regulate transcription, but a dimer of the same protein would bind more than twice as strongly, because its reaction with the palindromic base sequence would be cooperative, just like the binding of a dimeric antibody to the surface of a virus or bacterium. This makes for economy in size. Recognition can be doubly regulated: by the stereochemistry of the surfaces that DNA and protein monomers present to each other and by the matched spacings between the two halves of the palindrome and between the pair of protein surfaces that are complementary to it. The latter could also be varied by attaching the pair of protein monomers to a larger allosteric protein that changes its conformation in response to a hormone or a cytokine. In this manner binding of the protein pair to the DNA palindrome could be allosterically regulated.

A more direct mechanism operates in the *trp* repressor of *E. coli*. This is a dimer made up of two tightly linked helix-turn-helix motifs. When trytophan is bound to the repressor, two of its helices (one from each monomer) fit snugly into two neighboring major grooves of the operator DNA, but in the absence of trytophan the monomers collapse, the distance between the helices becomes too short, and two arginines essential for binding to DNA are buried inside the monomers. Yet another mechanism operates in the c-Jun and c-Fos proteins where DNA binding is regulated by oxido reduction of sulphydryl and phosphorylation of hydroxyl groups.

Not only can proteins change their conformation on binding to DNA; DNA can also change its conformation on binding to proteins. The most striking example of such a change occurs in the complex between the catabolic gene-activating protein (CAP) of *E. coli* and its DNA promoter sequence. CAP activates transcription of over 20 genes in response to its binding of adenosine-3′, 5′-monophosphate (cyclic or cAMP). It is a dimer of subunits each containing 209 amino acid residues. Each subunit is folded into two domains linked by a flexible hinge. The N-terminal domain contains the cAMP-binding site. The C-terminal domain contains the helix-turn-helix motif commonly found in prokaryotic DNA-binding proteins.

In the complex of CAP with a 30 base pair DNA promoter sequence, the C-terminal helices of the two protein subunits nestle in two adjacent major grooves of the DNA double helix. At each of these binding sites the double helix is bent by 40°, so that it wraps around the protein in a quarter circle. Binding of CAP may promote transcription by direct contact between it and RNA polymerase, and also by allowing the polymerase to make contact with nucleotide bases on either side of the CAP-binding site (Schultz et al., 1991).

In another complex between a prokaryotic transcription factor and its DNA operator, the *cro*-protein of *E. coli* phage λ, both protein and DNA change their structure. One cro monomer turns relative to its partner by 40° and the DNA bends by 40° into the shape of a boomerang (Brennan et al., 1990).

If we regard the nucleotide base sequence as a lock and the complementary protein surface as a key, we find that nature has evolved at least three different kinds of keys: the metal-free helix-turn-helix motif of the prokaryotic repressors and activators of transcription that has its counterpart in the eukaryotic homeo-domains; the former bind as dimers and the latter as monomers; the zinc-stabilized helix-turn-β-sheet motif of the eukaryotic zinc fingers; and the tandem β-turns of the SPXX motif. In the first two structures, the interaction between DNA and protein is centered around an α-helix fitting into a major groove. One would have thought that the angle of tilt of the α-helix would always follow the pitch of the DNA double helix, but in fact that angle varies widely in different protein-DNA complexes and seems to be determined by the pattern of hydrogen-bonding sidechains. Nor are α-helices essential for attaching proteins to the major groove of DNA; in the complex of the *met* repressor of *E. coli* with its operator, a β-strand fits into the major groove instead.

Detailed recognition relies predominantly on electrostatic interactions, that is, on hydrogen bonds between oppositely charged ions and/or opposed dipoles. There is a preference for amino acid sidechains that can form two hydrogen bonds such as arginines, glutamines, and asparagines. (Glutamic and aspartic acids would repel the DNA phosphates.) Hydrogen bonds go preferentially to G-C rather than A-T pairs, probably because the former are stronger dipoles, offer less steric hindrance, and can make pairs of hydrogen bonds on either side. (See Fig. A2.9 on page 293) This pattern of recognition is different from that between proteins where van der Waals interactions between non-polar surfaces play a larger role.

This chapter has dealt with proteins that recognize, and bind to, specific nucleotide base sequences in double-stranded DNA, because such proteins are prominent in oncogenesis. There are many other non-sequence-specific DNA-binding proteins, most important among them the histones that form the core of all eukaryotic chromosomes. They are basic proteins of five different kinds whose amino acid sequence has been largely conserved throughout evolution. One pair of each of four different histones is packed into an octamer. This octamer serves as a spool around which are wound 140 base pairs of double-helical DNA in two left-handed turns. The fifth histone ties together the two ends of the DNA that emerge from the spool. This assembly of histone and DNA is known as a *nucleosome*. Nucleosomes form a string of beads that is discernable under an electron microscope. In interphase chromosomes this string is coiled, first into a solenoid, and then into coils upon coils, like the coiffure of a rococo lady of fashion. This is how one meter of double helical DNA packs into the head of a single sperm (Kornberg and Klug, 1981).

X-ray analyses of crystalline nucleosomes and of the free histone octamer show it to be made up of two tightly packed dimers of two of the histones (H_3 and H_4) flanked on each side by a dimer of the other two (H2A and H2B). About half of each histone molecule is made of α-helices and a tenth of β-strands; they show up clearly in the electron density maps. The rest consists of invisible loops that may be disordered in the absence of DNA. The surface of the octamer forms a left-handed helical ramp tailored to fit the two turns of "superhelical" DNA (Richmond et al., 1984; Arents et al., 1991). On replication and transcription of DNA all the coils would have to be unwound. The mechanics of this process present an intriguing problem in topology.

A protein with histone-like properties found in *E. coli* embraces the double helix with two pairs of twisted β-strands; the DNA-gyrase in *E. coli* grips the double helix in a circular channel when its two monomers join to form a dimer (Tanaka et al., 1984; Wrigley et al., 1991). It seems that non-sequence-specific DNA-binding proteins will be found to interact with DNA in many different ways.

Far less is known about the interaction of proteins with RNA, because only one structure of a protein-RNA complex, that of an amino-acyl tRNA synthetase with tRNA, had been solved by the end of 1991. Recognition of tRNAs by their specific amino-acyl synthetases is the crucial step in the translation of the genetic into the amino acid code. Each of the 20 synthetases recognizes only the tRNAs for a single one of the 20 amino acids. It catalyses the displacement of the β-γ–diphosphate of ATP by the α-carboxylate of the amino acid. It then transfers the activated amino acid to the 3′-terminal adenosine of the tRNA. The structures of several amino-acyl synthetases and tRNAs in isolation have been determined, but the only complex of the two solved so far is that of glutaminyl-tRNA synthetase with its cognate tRNA.

Sequence-specific recognition of DNA by proteins usually involves only a few closely packed nucleotide bases and a minor portion of the protein surface, like that of the helix-turn-helix motif. By contrast, the 544 residue synthetase and the 76 nucleotide tRNA are in contact over most of their lengths, and 14— or as many as 16— widely spread nucleotides, including the amino acid acceptor strand and the anticodon loop that lie at opposite ends of the tRNA molecule, may contribute to recognition (for details see Rould et al., 1991; for review see Steitz, 1991).

Further Reading

Berg, J. 1990. Zinc finger domains: Hypotheses and current knowledge. *Ann. Rev. Biophys. and Biophys. Chem.* **19**:405–422.

Berg, J. 1990. Zinc finger and other metal-binding domains. Elements for interactions between macromolecules. *J. Biol. Chem.* **265**:6513–6516.

Freemont, P. S., A. N. Lane, and M. R. Sanderson 1991. Structural aspects of protein-DNA interactions. *Biochem. J.* **278**:1–23.

Harrison, S. C., 1991. A structural classification of taxonomy of DNA-binding domains. *Nature* **353**:714–719.

Harrison, S.C., and A. K. Aggarwal 1990. DNA recognition by proteins with the helix-turn-helix motif. *Ann. Rev. Biochem.* **59**:933–970.

McKnight, S. L. 1991. Molecular zippers in gene regulation. *Sci. Am.* **264 (April)**:32–39.

Najai, K. 1992. RNA-protein interactions. *Curr. Opinion Biol.* **2**:131–138.

Sauer, R. T., S. R. Jordan, and C. O. Pabo 1990. λ-repressor: a model system for understanding protein-DNA interactions and protein stability. *Adv. Protein Chem.* **40**:1–61.

Steitz, T. A., 1990. Structural studies of protein-nucleic acid interaction: the sources of sequence-specific binding. *Quart. Rev. Biophys.* **23**: 205–280.

Story, R. M., I. T. Weber, and T. A. Steitz 1992. The structure of the *E. coli rec A* protein monomer and polymer. *Nature* **355**:318–324. A new structure important for understanding the molecular mechanisms of genetic recombination and DNA repair.

Travers, A., 1989. DNA conformation and protein binding. *Ann. Rev. Biochem.* **58**:427–452.

4

How Drugs Recognize Proteins and Nucleic Acids

Proteins that alter their structure in response to chemical stimuli are called *allosteric*. Hemoglobin is the prototype of an allosteric protein, because it alters its structure on combination with, and release of, oxygen. This chapter begins with a piece of research where drugs found their own binding sites in hemoglobin and influenced the equilibrium between its two alternative structures in unexpected ways. The mapping out of these binding sites in atomic detail has led to some general rules about the stereochemistry of interaction between drugs and proteins that may be helpful in the design of drugs for binding to proteins of unknown structure. Knowledge of hemoglobin structure has also made it possible to redesign it as a possible blood substitute. Finally this chapter deals with anticancer drugs that stop cell division by disturbing the structure of double-helical DNA.

Most drug targets are membrane-bound receptors that alter their structures in response to stimuli by neurotransmitters, hormones, cytokines, and other factors. None of them has yet been crystallized, and their structures are still unknown. Unless an accidental observation shows the way, the development of drugs against such targets can be a long and tortuous process.

Hemoglobin as Drug Receptor Model

Until 1980, the few complexes of drugs bound to proteins analyzed crystallographically included those of myoglobin and hemoglobin with the anesthetic gases xenon and CH_2Cl_2; of carbonic anhydrase with sulfanilamide (Lindskok et al., 1971); and of dihydrofolate reductase with methotrexate (Matthews et al., 1978). Xenon and

CH_2Cl_2 occupy crevices in the globin (Nunes and Schoenborn, 1973), while the other two drugs were analogues to the enzymes' natural substrates and acted as inhibitors at their active sites. In a search for agents that would alleviate sickle cell anemia or lower the oxygen affinity of hemoglobin, my colleagues and I have made X-ray studies of complexes of hemoglobin with several compounds, three of them commonly used drugs. The compounds were cocrystallized with human carbonmonoxy- or deoxyhemoglobin. All crystals were isomorphous with those of the drug-free hemoglobin, which made it easy to locate the drugs by X-ray analysis. Their binding sites are diverse and coincide neither with the active site at the heme nor with the site of the natural allosteric effector 2,3-diphosphoglycerate (DPG) (Perutz et al., 1986).

Hemoglobin is in equilibrium between two alternative structures: the deoxy- or T-structure with low and the oxy- or R-structure with high oxygen affinity. It is a tetramer made up of two pairs of polypeptide chains, called α and β, each harboring one heme. The tetramer has twofold symmetry. A water-filled cavity or channel passes through the center of the molecule along the twofold axis (Fig. 4.1). This channel is wider in the T- than in the R-structure. In nature, the oxygen affinity of hemoglobin is lowered in a physiologically beneficial way by chloride ion, CO_2, H^+, and 2,3-diphosphoglycerate. These are known as allosteric effectors, and they exert their influence by binding to the T-structure more strongly than to the R-structure. Their binding generates hydrogen bonds that stabilize the T-structure, which in turn lowers the oxygen affinity, expressed as P_{50}, the partial pressure of oxygen needed to saturate half the hemes with oxygen. 2,3-diphosphoglycerate binds in the water-filled cavity to a complementary constellation of cationic groups between the two β-chains that exists only in the T-structure where the channel is wide. It must dissociate before the channel can narrow on transition to the R-structure (Perutz, 1990).

Clofibric acid, a commonly used antilipidemic drug, has anti-sickling properties and also lowers the oxygen affinity of hemoglobin (Formulae 4.1, page 122). X-ray analysis showed that two pairs of clofibric acid molecules bind to the central channel of the T-structure with high occupancy and another pair binds there with weaker occupancy; in the R-structure the channel is too narrow to accommodate the drug. Instead, one pair of clofibric acid molecules binds in a cleft between two helices that is normally occupied by the indol side chain of a tryptophan (Abraham et al., 1983).

Figure 4.1. α-carbon skeleton of human hemoglobin viewed along its twofold symmetry axis, showing the four hemes and the two drug molecules LR16 in the central, water-filled cavity, indicated by arrows.

The antilipidemic drug bezafibrate resembles two clofibric acid groups in tandem. It lowers the oxygen affinity more strongly than clofibric acid, but, disappointingly, it promotes sickling rather than inhibiting it (Perutz and Poyart 1983). Nevertheless, agents that lower the oxygen affinity would be clinically useful, for example, for improving oxygen delivery to infarcted tissue, to tumors before irradiation, or after shock. Our finding therefore stimulated a search for other compounds with similar and preferably stronger action. Table 4.1 lists these compounds together with their influence on the oxygen affinity. (See Formulae 4.1)

As a first step bezafibrate was replaced by a compound in which the $-CO-NH-CH_2-CH_2-$ bridge between the two benzene rings was replaced by a urea moiety, and an additional chlorine was attached to the first benzene in position 3 (LR16). This derivative lowered

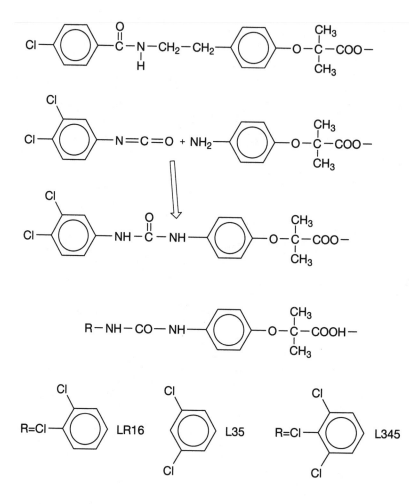

Bezafibrate

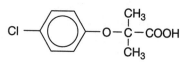

Clofibric acid

Formulae 4.1.

Table 4.1. Oxygen Binding Parameters for Fresh Normal Red Cell Suspensions at 37°C[a]

EFFECTOR (mM)	P_{50} (mmHg)	n_{50}	P_{50}EFF/P_{50}CONTROL	SLOPE
none (control)	23.4	2.34		
Bzf				
1.0	28.6	2.0	1.2	0.2
5.0	37.6	2.3	1.6	
LR16				
0.2	28.3	2.0	1.2	
0.5	48.6	2.2	2.0	0.57
1.0	69.5	2.1	3.0	
L35				
0.1	28.8	2.0	0.2	
0.2	32.6	2.2	1.82	0.63
0.5	69.6	2.1	3.0	
1.0	123.0[b]	2.0	5.3	
L345				
0.1	33.5	2.0	1.4	
0.2	58.0	1.8	2.15	0.98
0.5	160.0[b]	1.6	11.0	

[a]Conditions: 0.14 M NaCl, 50mM Bis–Tris (pH <7.5) or Tris–HCl buffer (pH >7.5), 37°C. P_{50}eff/P_{50}control is the ratio of P_{50} with the effector to the control without it. The slopes were calculated as the difference in log P_{50} for a given variation of log (effector) concentration in molar.
[b]These curves were recorded at pH 7.8 to ensure full oxygenation of the red cells under 1 atm of oxygen. P_{50} values were corrected to pH 7.4 by using a Bohr factor of 0.60.

the oxygen affinity four times more strongly than bezafibrate. An even stronger effect was achieved when the two chlorines were placed in the 3,5 rather than the 3,4 positions (L35); finally, the 3,4,5-trichloro derivative (L345) had the strongest influence of all (Lalezari et al., 1988, 1990).

X-ray analysis allowed the stereochemistry of the interaction of each of these compounds with the globin to be determined. Bezafibrate binds to a single pair of sites in the central channel, adhering mainly to the α-chains. The contacts between it and the globin include one strong electrostatic interaction between the carboxylate of the drug and the guanidinium group of an arginine, as well as some purely nonpolar interactions and some weakly polar ones. The latter were explored in detail (Fig. 4.2). p-chlorobenzene has a dipole moment of 1.57 debyes, with a negative pole at the chlorine. In bezafibrate this chlorine is in contact with the

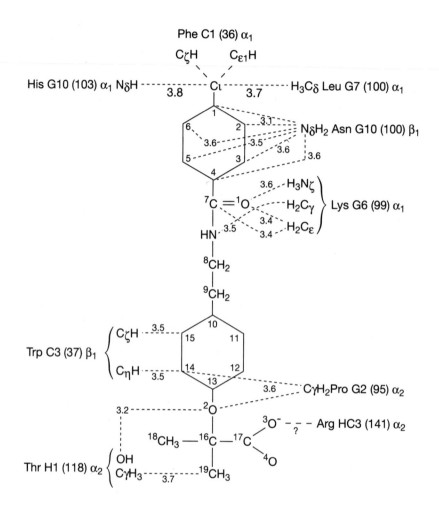

Bezafibrate in central cavity

Figure 4.2. Stereochemistry of interactions between bezafibrate and the surrounding globin. Note the interactions between the electronegative chlorine and the electropositive hydrogens of its surrounding hydrocarbons. Note also the interactions between the amino group of Asn 108 and the π-electrons of the chlorobenzene ring. (Reproduced, by permission, from Perutz et al., 1986)

fractional positive charges on the hydrogens of adjacent sidechains: a histidine, a phenylalanine, and a leucine. The amide NH_2 of an asparagine sidechain interacts with the drug's benzene ring. The amide has a dipole moment of 3.4 debyes with its positive pole on the NH_2; this appears to form a hydrogen bond with the π-electrons of the benzene, with a bond energy of about half that of an NH···OC bond. Similar hydrogen bonds between amides and aromatic rings were later found to contribute to the stability of many proteins (Levitt and Perutz 1988; Burley and Petsko, 1986).

LR16 bound to the same pair of sites as bezafibrate. Its fourfold stronger influence on the oxygen affinity appears to be due partly to the hydrophobic effect of the additional chlorine; partly to the larger dipole moment of the dichlorobenzene ring which strengthens the interactions with the fractional positive charges on the surrounding hydrogens; and partly to an entropic effect. In bezafibrate, both benzene rings are free to rotate about four bonds in the bridging aliphatic chain. In LR16, L3,5, and L3,4,5, on the other hand, the chlorobenzene ring is coplanar with the bridging urea moiety, because electron repulsion by the chlorines gives the C_6–NH bond partial double-bond character. The second benzene ring can rotate only about the HN-C_6 bond. On binding to the globin, these three derivatives therefore lose less rotational entropy than bezafibrate, which raises their binding energy (Lalezari et al., 1988).

L3,5 and L3,4,5 lower the oxygen affinity more strongly than LR16 because each of them binds to an additional pair of sites in the central cavity that are only weakly occupied by clofibrate and bezafibrate. That binding is due mainly to hydrogen bonds between the NH groups of their urea moieties and the carboxylate of a glutamate. The positions of the second pair of sites occupied by L3,5 differ from those occupied by L3,4,5, even though L3,4,5 differs from L3,5 only by the attachment of one more chlorine (Fig. 4.3). The dichlorobenzenes of L3,5 stack in pairs, with the chlorines staggered relative to each other, while the four trichlorobenzenes of L3,4,5 are piled together in one closely packed stack in which some of the chlorines are eclipsed.

These and other studies have led to some generalizations concerning the interactions between drugs and proteins.

1. The greater the hydrophobic effect, that is, the greater the number of water molecules set free on binding of the drug to the protein with a consequent gain in entropy, the greater the affinity of the drug for the protein.

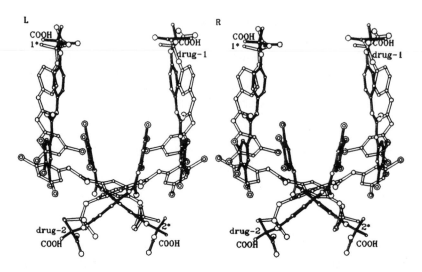

Figure 4.3. Stereodrawing of the relative positions of the drugs L3,5 (open bonds) and L3,4,5 (closed bonds) in the central cavity of hemoglobin. The vertical molecules occupy the same sites as bezafibrate. (Reproduced, by permission, from Lalezari et al., 1990)

2. The smaller the loss of internal rotational entropy of the drug on binding to the protein, the greater the affinity.
3. Drugs bind in niches of the protein that fit their van der Waals volume and take up positions that maximize the mutual polarizabilities at their contacts with the protein. Electrostatic interactions between drug and protein cover a wide range of polarities, from strong hydrogen bonds between ionized groups of opposite charge, to weak interactions between aromatic quadrupoles and nonpolar interactions between aliphatic hydrocarbons.
4. Small stereochemical changes in the drug (for example, the shift of one chlorine from an ortho to a meta position) can open additional binding sites or make the drug move from one binding site to another. Such changes are hard to predict.
5. Drugs may influence the allosteric equilibrium of a protein receptor in the same direction as the natural effector (2,3-diphosphoglycerate in hemoglobin), even if they are chemically unrelated to it, because proteins may offer a variety of binding sites that are not used in nature. Most drug

receptors are membrane proteins that change their structure in response to chemical stimuli; in other words, they are allosteric. Our experience with hemoglobin implies that a drug may influence the allosteric equilibrium of receptors in the same direction as the natural hormone or transmitter, even though the drug is chemically unrelated to it and binds to a different site or sites (Perutz et al., 1986).

4 and 5 are important new generalizations that have emerged from this work, which proved that there can be a spin-off from applied research to basic science, the reverse of the usual.

Binding of Cytostatic Drugs to DNA

Watson and Crick's double-helical model of DNA is the key to our understanding of the action of many mutagens, carcinogens, anti-tumor agents, and antibiotics. These compounds disturb the double helix in a variety of ways: by either breaking or cross-linking the phosphate ester chains, by cross-linking nucleotide bases, by intercalation between the base pairs, or by binding to one of the grooves of the double helix (Fig. 4.4) (Waring 1981).

Acridines are powerful mutagens that are also used as anti-tumor agents and disinfectants. In 1961 Lerman discovered that one of these, proflavine, raises the viscosity of solutions of calf thymus DNA and lowers its sedimentation constant, which suggested that proflavine lengthens the DNA. Lerman showed that these effects could be accounted for if the polycyclic ring of proflavine intercalated between the base pairs of the double helix (Lerman, 1961). His discovery had far-reaching consequences, because it suggested that intercalation of proflavine and other acridines shifts the reading frame of the genetic message. This idea led Crick, Brenner, and others to perform the experiments that established the triplet nature of the genetic code (Brenner et al., 1961; Crick et al., 1961).

Crystallographic proof had to wait until 1975, when Sobell and others solved the structure of the ribodinucleoside 5-iodo-uridyl (3′–5′) adenosine complexed with ethidium bromide (Formula 4.2). In the crystal two ribonucleosides form part of a double-helical turn, with the two base pairs stacked in parallel 6.8Å apart (Fig. 4.5). Midway between them lies the phenanthridinium ring of ethidium bromide whose ethyl and phenyl sidechains protrude into the narrow groove (Tsai et al., 1975; Sobell et al., 1977).

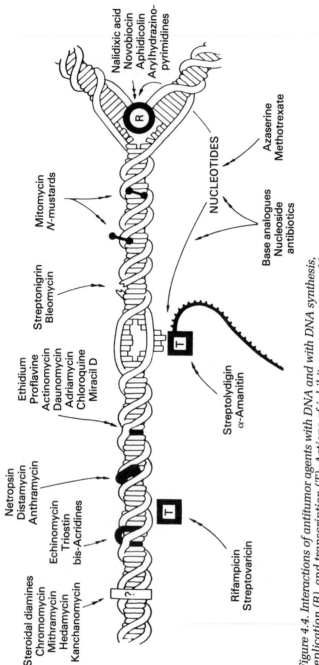

Figure 4.4. Interactions of antitumor agents with DNA and with DNA synthesis, replication (R), and transcription (T). Actions of inhibitors are represented by double-headed arrows and are purely diagrammatic without any implication that the sites of action represented here account for all the modes of action of the drugs. (Reproduced, by permission, from Gale et al., 1981)

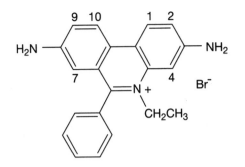

Ethidium bromide

Formula 4.2.

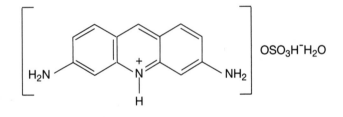

Proflavine sulfate

Formula 4.3.

Figure 4.5. Stereodrawing of the intercalation of ethidium between the base pairs of the dinucleoside monophosphate 5-iodouridylyl (3′–5′) adenosine (iodo UpA). (Reproduced, by permission, from Tsai et al., 1977)

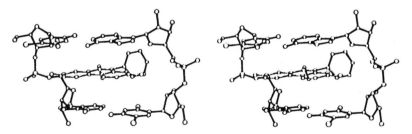

Neidle and others determined the structures of the complexes of proflavine with base-paired ribo- and deoxyribodinucleosides (Formula 4.3). When proflavine intercalates between the base pairs of CpG or dCpG, the distance between the base pairs expands from 3.4 to 6.8Å as in Sobell's ethidium bromide complex. The long axis of the proflavine lies approximately normal to the direction of the hydrogen bonds between the bases. In the ribose complex the amino groups of proflavine form hydrogen bonds with O'_2 of opposite ribose sugars (Neidle et al., 1977; Shieh et al., 1980).

Actinomycin D is a fungal antibiotic that stops the growth of rapidly dividing cells and is used, in combination with other agents, for the treatment of children's sarcomas (Formula 4.4). Actinomycin D binds tightly to double-helical DNA and inhibits

Formula 4.4.

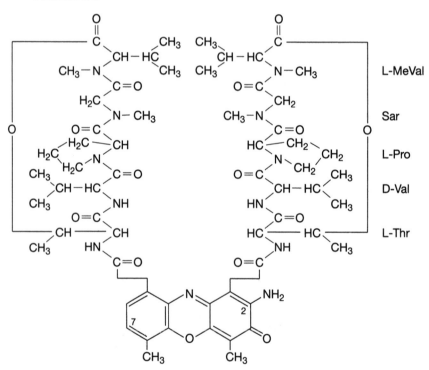

Me Val, methyl-valine; Sar, sarcosine; Pro, proline; Val, valine; Thr, threonine

its transcription into messenger RNA. It consists of two cyclic pentapeptides attached to a phenoxazone ring and binds specifically

to $\overset{\text{G}\cdots\text{C}}{\underset{\text{C}\cdots\text{G}}{}}$ pairs. Jain and Sobell have determined the crystal structure of a complex of one molecule of actinomycin with two molecules of deoxyguanosine (Jain and Sobell, 1972). They found the phenoxazone ring sandwiched between two guanine rings to which the peptides were linked by hydrogen bonds. These explained why actinomycin D recognizes specifically two opposed G–C pairs stacked above each other.

Daunomycin and doxorubicin are two antitumor drugs that differ by only one OH group; the former is used to treat leukemia and the latter to treat breast, colon, and other cancers (Formula 4.5). The drugs consist of an aglycon chromophore that is part of an anthracyclin skeleton, coupled to an amino sugar that carries a positive charge; they inhibit both replication and transcription of DNA. When combined with the hexanucleotides (CGATCG) or d(CGATCG), the anthracyclin moiety intercalates between adjacent G–C pairs; ring A and the amino sugar protrude into the minor groove and ring D into the major groove. The ammonium group of the sugar makes hydrogen bonds with phosphate oxygens of the DNA. The double helix accommodates the intercalated aromatic rings by a purely local unwinding, consisting of a reduction in the twist angle between successive nucleotides from $35°$ to $30°$ (Fig. 4.6, page 132) (Wang et al., 1987; Moore et al., 1989). Nogalamycin is another anthracycline antibiotic that intercalates between the base pairs of DNA and also forms hydrogen bonds with specific bases (William et al., 1990).

It is not clear why daunorubicin and doxorubicin inhibit DNA replication and transcription. The binding of the sugar with its strong hydrogen bonds to the phosphate oxygens may inhibit the binding of the necessary enzymes to the DNA, but an assay with nicked circular DNA and T4 phage ligase showed no correlation between the ability of various anthracyclins to alter DNA structure and their cytotoxic and antitumor activity, which suggests that other effects may also be involved.

Triostin A and echinomycin are cyclic depsipeptides carrying two colored quinoxalin rings (Formula 4.6). They bind to the minor groove of double-helical DNA with their chromophores aligned to inter-calate so that they bracket two adjacent base pairs (Quigley et al., 1986).

Distamycin, netropsin, and Hoechst 33258 constitute another group of antiviral, antitumor drugs that contain chains of heterocyclic rings linked by amide groups (Formula 4.7). The

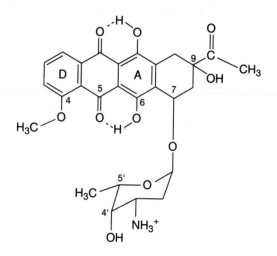

Daunomycin

INTERCALATORS

Formula 4.5.

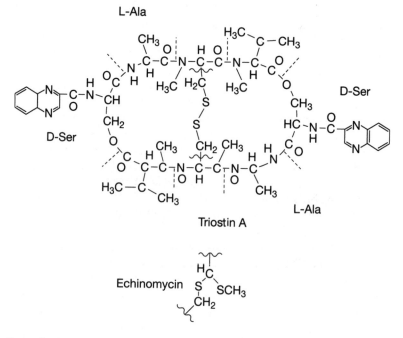

Triostin A

Echinomycin

Formula 4.6.

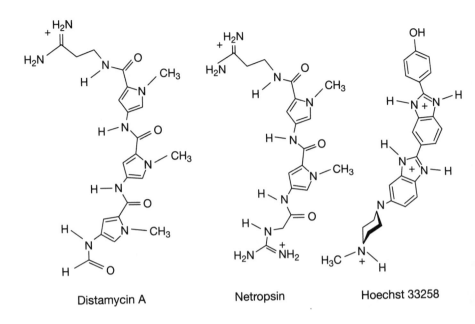

Distamycin A   Netropsin   Hoechst 33258

MINOR GROOVE BINDERS

Formula 4.7.

amide groups impose a curvature on the chains that makes them complementary to a sequence of A–T base pairs in the minor groove of double-helical DNA. They replace the water molecules normally bound there (Fig. 4.7). These compounds are too toxic for clinical use, but they serve as probes for drug resistance of tumor cells (Kopka et al., 1985).

Cis-platin and carboplatin have made most testicular cancers curable and are two of the most successful antitumor drugs known (Formula 4.8). They react with adjacent guanines in a double helix, with the elimination of two molecules of HCl or of cyclobutane-1,1 dicarboxylate, and by formation of covalent bonds between platinum and N_7 of the two guanosines. The crystal structure of the complex of cis[Pt(NH$_3$)$_2${d(pGpG)}] shows the planes of the two coordinated guanine bases to lie nearly at right angles to each other (Fig. 4.8). This is a necessary consequence of the planarity of the (PtN$_4$) complex and of the $\frac{C}{C}{>}$N–Pt groups. Rotation is allowed only around the Pt–N bonds, and steric hindrance between the

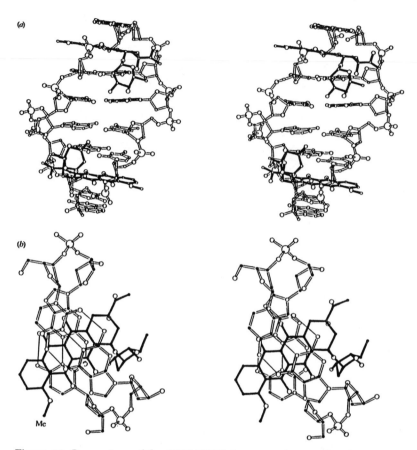

Figure 4.6. Stereoviews of the d(CGATCG)-daunomycin complex. (A) View of the minor groove side of the complex. The drug is depicted with dark bonds. Note how the daunosamine moiety covers a large section of the minor groove. (B) A view perpendicular to the drug's chromophore including the CpG intercalation site. The drug is shown with dark bonds. This diagram illustrates the head-on intercalation. (Reproduced, by permission, from Moore et al., 1989)

(NH$_3$) groups coordinated to the platinum and O$_6$ of each guanine restricts the allowed dihedral angles between the guanine and the (PtN$_4$) planes to values larger than 60°. It can be shown to follow that the dihedral angle between the guanines must be at least 76°. Such a large angle between the planes of two bases that should be nearly parallel to each other disrupts the double-helical structure of DNA and disturbs both replication and transcription. This

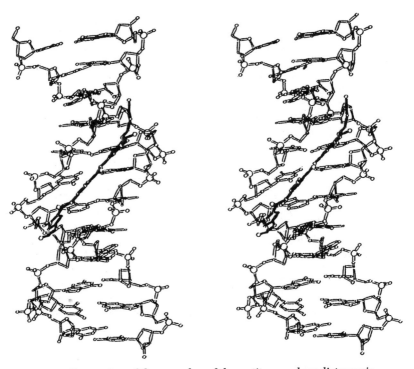

Figure 4.7. Stereoview of the complex of the antitumor drug distamycin with deoxy (CGCAAATTTGCG)₂. The drug binds to the minor groove and its NH groups form hydrogen bonds with the A–T pairs of the DNA, displacing the water molecules normally bound there. (Reproduced, by permission, from Coll et al., 1989)

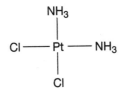

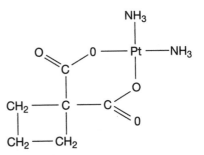

Formula 4.8.

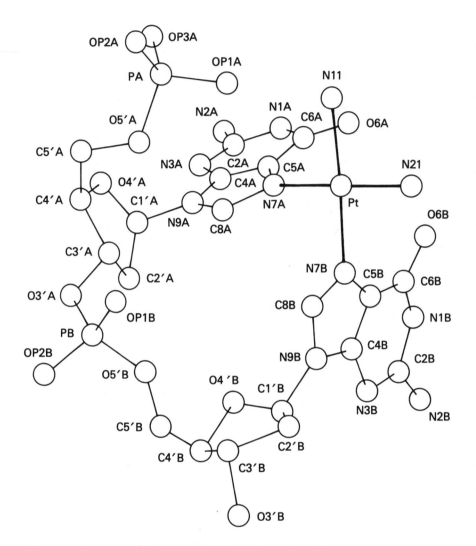

Figure 4.8. Structure of cis-[Pt(NH₃)₂(pGpG). (Reproduced, by permission, from Sherman et al., 1988)

disturbance together with the blocking of the two guanines must be responsible for the drugs' pharmacological action (Sherman et al., 1985; Sherman et al., 1988; Sundquist et al., 1987).

In aqueous solution water molecules replace the chlorines in cis-platin, and [cis-Pt(NH₃)₂Cl(H₂O)]⁺ may be the species that reacts

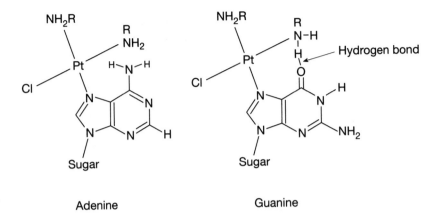

Figure 4.9. Difference between binding of cis-platin to adenine and guanine.

with DNA *in vivo*. The severe side effects of cis-platin may be due to its reaction with sulfhydril groups and to the reaction of its aquo-compounds with guanosine-5′-monophosphate (Djuran et al., 1991). Its preferred reaction with guanine rather than adenine bases in DNA arises because its amino group can form a hydrogen bond with the CO of guanine, but not with the NH_2 of adenine, as shown in Figure 4.9. (For review, see Reedijk et al., 1987; Van der Veer and Reedijk, 1988).

Further Reading

Gao, X. and D. J. Patel 1989. Antitumor drug–DNA interactions: NMR studies of echinomycin and chromomycin complexes. *Quart. Rev. Biophys.* **22**:93–138.

Hol, W. G. J. 1987. Applying knowledge of protein structure and function. *Trends in Biotechnology* **5**:137–143.

Kennard, O. and W. N. Hunter 1989. Oligonucleotide structure: A decade of results from single crystal X-ray diffraction studies. *Quart. Rev. Biophys.* **22**:327–380.

Lippard, S. J. 1987. Chemistry and molecular biology of platinum anticancer drugs. *Pure Appl. Chem.* **59**:731–742.

Reedijk, J. 1987. The mechanism of action of platinum antitumor drugs. *Pure Appl. Chem.* **59**:181–192.

Saenger, W. 1984. *Principles of Nucleic Structure.* Springer-Verlag, New York.

5

Drugs Made to Measure: From the First X-Ray Diffraction Picture of a Pepsin Crystal to Drugs Against HIV and Hypertension

The discovery of new drugs often originated from accidental observations, like Alexander Fleming's discovery of the first antibiotic, penicillin:

> While working with staphylococcus variants, culture plates were set aside on the laboratory bench and examined from time to time. In the examination these plates were necessarily exposed to the air and became contaminated with various micro-organisms. It was noticed [when Fleming returned from his summer vacation] that around a large colony of contaminating mould the staphylococcus colonies became transparent and were obviously undergoing lysis.

These three momentous sentences were the crux of Fleming's classic paper, "On the Antibacterial Action of Cultures of a Penicillium with Special Reference to Their Use in the Isolation of B. Influenza" (Fleming, 1929).

Fleming had no means of finding out how penicillin works because drug receptors were black boxes and drug designers had to grope in the dark. Protein crystallography has opened a new submicroscopic world where potential drug targets can be mapped out in atomic detail. These targets are enzymes, and the desired drugs are inhibitors of these enzymes. This chapter shows how Bernal and Crowfoot's original observation of X-ray diffraction patterns from crystals of pepsin led 55 years later to the design of inhibitors of the HIV proteinase and of one of the enzymes that controls blood pressure. It recalls how one of the first enzyme structures to be solved, that of pancreatic carboxypeptidase A, led, in a way that could never have been foreseen when W. N. Lipscomb started the work in 1961, to the design of potent inhibitors of another enzyme that controls blood pressure. Finally, I shall describe how the recent solutions of the structures of two enzymes from *E. coli* have led to the mapping of a new potential target for drugs against HIV.

To design effective inhibitors, one must know not only what their targets look like, but also how they work. I shall therefore go into the details of both the structures and the catalytic mechanisms of the enzymes concerned—not least because, as a chemist, I find them both fascinating.

Acid Proteases and HIV

The rich X-ray diffraction patterns of protein crystals were first discovered in the acid proteinase pepsin, but all attempts to solve the structure of pepsin failed until 1977 when the groups of M. N. C. James in Edmonton, Alberta, and Tom Blundell in London crystallized fungal acid proteinases that are homologous to pepsin and proved more amenable to X-ray analysis (Hsu et al., 1977; Subramanian et al., 1977). After that Natalia Andreeva in Moscow solved the structure of porcine pepsin by analogy with the structures of the fungal enzymes (Andreeva et al., 1978). A refined structure of porcine pepsin has only just been published (Sielecki et al., 1990a). All these acid proteinases consist of two lobes separated by a cleft that contains the catalytic site. That site is made up of two aspartate residues, one on either wall of the cleft, and a water molecule bound between them. The entrance to the active site cleft is controlled by long mobile loops of β-strands protruding from either lobe; they form flaps capable of closing in on the substrate and binding to it (Figs. 5.1 and 5.2). Until AIDS confronted us, this seemed merely another piece of recondite knowledge to be thrust down the throats of hapless medical students who are already saturated with useless facts.

Several proteins of the human immunodeficiency virus (HIV) are synthesized as a single polypeptide chain that an acid proteinase then splits into its components. The amino acid sequence of that proteinase, deduced from the nucleotide sequence of its gene, suggested that its structure is similar to that of the acid proteinases of the pepsin family. Three separate groups of crystallographers (two in the United States and one in London) set out to crystallize the protein and determine its structure. It was indeed found to be

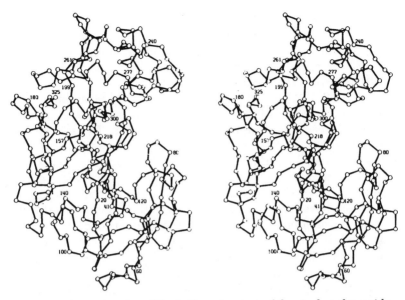

Figure 5.1. Stereodrawing of the tertiary structure of the single polypeptide chain of rhizopus pepsin. The catalytic site lies in the large cleft. (Reproduced, by permission, from Suguna, et al., 1987)

Figure 5.2. Pair of aspartates forming the catalytic sites of pepsin and other acid proteases. The interatomic distances are those of porcine pepsin taken from Sielecki et al. (1990a). The pK_as of the aspartates are lowered by hydrogen bonds donated by two main chain NHs and by the OH of a serine and a threonine. (Reproduced, by permission, from Suguna et al., 1987)

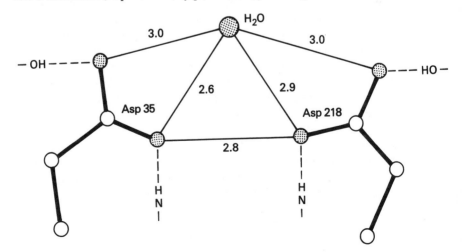

pepsin-like, with the important difference that the two separate lobes in the single polypeptide chain of pepsin are replaced by two identical chains of half the length, each contributing one lobe with one aspartate to the active site (Fig. 5.3) (Navia et al., 1989; Wlodawer et al., 1989; Lapetto et al., 1989).

An inhibitor of the HIV acid proteinase could be a potent drug against AIDS, especially in combination with other drugs such as the RNA transcriptase inhibitors already in use. The structure of the proteinase opened the possibility of making such inhibitors to measure. M. V. Toth and others have made an oligopeptide whose central amide bond is replaced by a $-CH_2-NH-$ that the proteinase cannot split. A. Wlodawer and his colleagues and also Fitzgerald and others have found a piece of information that is vital for the design of even more potent inhibitors: combination of the enzyme with the inhibitor makes the two peptide loops at the entrance to the active site close in on it (Fig. 5.4) (Miller et al., 1989; Fitzgerald et al., 1990). Compounds related to the oligopeptide of Toth and others are in clinical trial as of 1991, and new inhibitors are being designed to immobilize the gate to the active site.

Desjarlais and others have used the structure of the HIV protease for a new approach to drug design (Desjarlais et al., 1990). Traditionally, the search for a new drug is begun by the random screening of many thousands of organic molecules. When an active compound is found, derivatives are synthesized and screened in order to map out stereochemical relationships between constitution and activity from which the structure of the active site can be inferred. Desjarlais and others have developed computer programs that allow this cumbersome procedure to be inverted. Using Wlodawer et al.'s atomic coordinates of the inhibitor-free HIV-1 protease, they generated a 24Å long and 8Å wide negative image of its active site. They then searched the Cambridge Structural Database for organic compounds that fit it and formed hydrogen bonds with it. Their search homed in on bromoperidol, a member of a class of compounds that serve as antipsychotic agents (Formula 5.1 on page 145). Bromperidol inhibited the HIV-1 protease with an $I_{50} = 25$mM.[1] This level of activity is several orders of magnitude too low for clinical use, but bromoperidol can now serve as a lead compound for the synthesis of more potent derivatives. Knowledge of the enzyme structure combined with ingenious

---

[1]I_{50}: inhibitor concentration that halves the enzyme's activity.

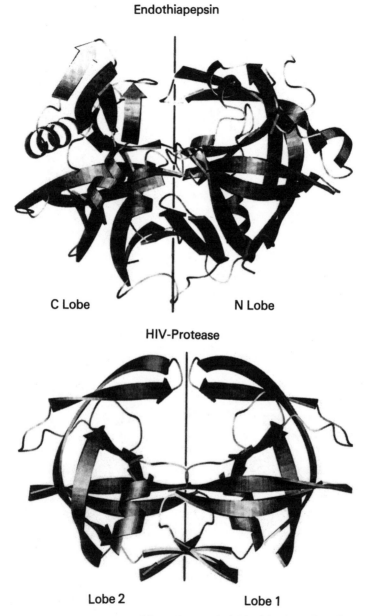

Endothiapepsin

C Lobe   N Lobe

HIV-Protease

Lobe 2   Lobe 1

Figure 5.3. Comparison of the polypeptide fold of the fungal acid protease endothiapepsin, which is similar to that of mammalian and rhizopus pepsin, with that of the HIV protease. (Courtesy of Dr. Jon Cooper, Birkbeck College, London)

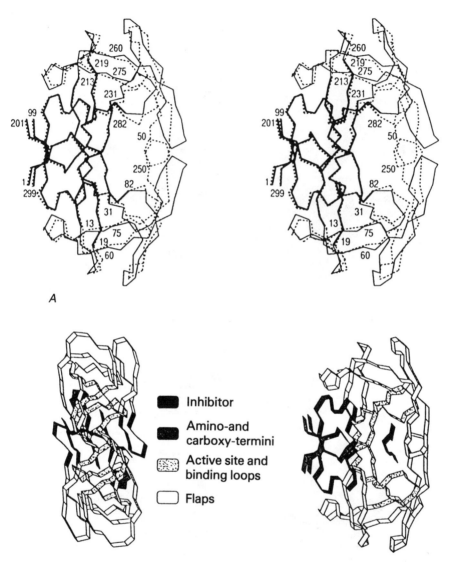

A

B

Figure 5.4. (A) Stereodrawing of the free HIV protease (full lines), superimposed on that of the protease bound to the inhibitor acetyl-pepstatin (dotted line). (The inhibitor is omitted.) (Reproduced, by permission, from Fitzgerald, et al., 1990) (B) Diagrammatic drawings of the protease with the inhibitor bound. (Right) Viewed from the same direction as in (A). (Left) Viewed at right angles to that direction. (Reproduced, by permission, from Miller et al., 1985)

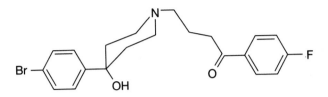

Formula 5.1. Bromoperidol.

programs led the authors to its discovery in just over 10 hours of computing time, compared to the months or years often required for the discovery of a lead compound by classical methods (Goodford, 1985).

Catalytic Mechanism of Pepsin

Pepsin is most active at pH 2.0, the pH in the stomach, at which most other proteins are denatured. Moreover, it carries a net negative charge even at a pH of 1.0 and has an isoelectric point below 1.0. It was a triumph of structure analysis to explain these remarkable properties.

When surrounded by water, external carboxylates of proteins normally have pK_as between 3.6 and 4.5, so that one would expect them to be protonated and discharged at pH 1.0. On the other hand, internal carboxylates can have their pK_as shifted, sometimes by several units, by the electric fields of neighboring protein atoms or ions. Their pk_as may be raised by neighboring anions and lowered by neighboring cations or by hydrogen bond donors such as NH and OH, which repel other protons.

Porcine pepsin, which was the first mamalian acid proteinase whose structure was accurately known, contains 29 aspartates and 13 glutamates, but only two arginines, one lysine and one histidine in its 326 amino acid–long polypeptide chain. Eleven of the aspartates and glutamates are buried, either wholly or partly, two of them in the active site cleft. The carboxylates of the two aspartates in that cleft are hydrogen-bonded to each other, which lowers the pk_a of the proton donor to 1.2 and raises that of the acceptor to 4.7. Both carboxylates accept hydrogen bonds from the neighboring water molecule and from NHs and OHs of the protein (Fig. 5.2). One of the other buried aspartates has its carboxylate hydrogen-bonded to an arginine, and another is close enough to an arginine to have its negative charge partly compensated. The remainder accept hydrogen bonds from NHs, OHs, or buried water molecules that repel protons and therefore stabilize their negative charges (Sielecki et al., 1990a). $NH \cdots O^-$ and $O \cdots HO^-$ bonds can be very strong; they may provide enough energy to keep the structure of pepsin intact at low pHs.

Pepsin's long active site cleft is designed to accommodate the unfolded polypeptide chains of proteins that are denatured in the acid environment of the stomach. The lining of the cleft ensures that the chains arrange themselves so that peptide bonds between large hydrophobic residues face the catalytic site. The enzyme then cleaves these bonds. Suguna and others (1987) have tried to infer the catalytic mechanism from the structure of one of the fungal acid proteinases which they combined with a pseudosubstrate, an octa-peptide that had one of its peptide carbonyls reduced to CH_2. This inhibitor lined the active site cleft with that reduced bond close to the aspartates. Figure 5.5 shows the proposed catalytic mechanism in outline. The partial double bond character of the peptide bond confines the atoms Cα-NH-CO-Cα to a plane. Proteolytic enzymes induce the transition state by distorting the coordination around the carbonyl carbon to a nearer tetrahedral one in which the carbon and nitrogen are linked by only a single bond. In pepsin, the bound water molecule between the two aspartates probably pushes the carbonyl oxygen of the scissile bond away from the plane of the peptide and toward its position in a tetrahedral intermediate. That distortion would make the carbonyl carbon more susceptible to nucleophilic attack by the water, shown in step 2.[1] Next, one of the water protons would discharge aspartate 35, whose negative charge would pass to the tetrahedral intermediate (step 3). That transfer of charge would facilitate the jumping of the proton from aspartate 35 to the peptide nitrogen, followed by cleavage of the peptide bond and dissociation of the products in steps 3 and 4. The authors explain how, in addition, strategically placed polar groups of the enzyme would help to polarize the scissile bond in the right direction for catalysis at each of these steps.

The constellation of the aspartates and the water molecule is the same in all acid proteases. The enzymes owe their versatility and specificity to the variable lining of their clefts. Renin, for example, cleaves only the bond between leucine and valine ten residues from the amino

---

[1]Nucleophilic attack means the reaction of an electron-rich atomic region, such as the lone-pair electrons of an oxygen atom, with an atom carrying a partial positive charge, such as a carbonyl carbon.

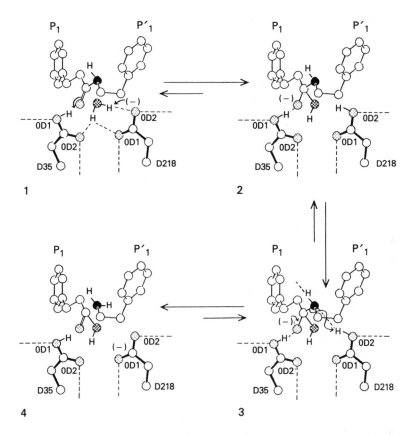

Figure 5.5. Catalytic mechanism of pepsin proposed by Suguna, Padlan, Smith, Carlson, and Davies. 1, 2, 3, and 4 refer to the catalytic steps described in this text. (Reproduced, by permission, from Suguna et al., 1987)

amino end of angiotensinogen, the precursor of the hormone angiotensin; it is specific for that one particular amino acid sequence.

Development of Antihypertensive Drugs

Angiotensin II constricts blood vessels and stimulates the secretion of aldosterone which increases the retention of sodium ions and water by the kidneys. Its precursor is the protein angiotensinogen. The acid proteinase renin that is secreted by the kidneys splits the inactive decapeptide angiotensin I from the N-terminus of angiotensinogen. The angiotensin-converting enzyme then cleaves the C-terminal histidyl-leucine from angiotensin I and thus converts it into the active octapeptide angiotensin II (Formula 5.2). The structure of the angiotensin-converting enzyme is still unknown, and that of renin has been solved only in 1990. The design of inhibitors therefore has had to be based on their structurally known analogues: pepsin and the fungal acid proteases for renin, and the pancreatic carboxypeptidase A and thermolysin for the angiotensin-converting enzyme. The gene for the angiotensin II receptor itself has now been cloned. The amino acid sequence deduced from it is homologous to those of a family of receptors that contain seven hydrophobic segments indicative of transmembrane helices (Sasaki et al., 1991). (See Chapter 11.) The gene could now be introduced into single cells, such as frog oocytes, capable of expressing the receptor. Once this has been achieved, antagonists could be assayed in such cells.

Design of Renin Inhibitors

Only trace quantities of renin or its precursor, prorenin, can be extracted from kidneys. Human prorenin was therefore synthesized in cultures of Chinese hamster ovary cells transfected with human prorenin c-DNA and converted to renin by cleavage with immobilized trypsin. The structure of this recombinant human renin bears a close resemblance to that of porcine and, by implication, human pepsin (Sielecki et al., 1990a,b). Its single polypeptide chain contains 327 amino acid residues compared to pepsin's 325, and 131 of these are identical in the two proteins. The mean difference between 279 well-determined α-carbon positions in the two enzymes is only 1.33Å, and their similarity is even closer near the two catalytic aspartates in the active site (Fig. 5.6). These similarities conceal profound differences between these enzymes' specificities and pH profiles. Pepsin preferably cleaves peptide bonds between leucine, phenylalanine, tryptophan, or glutamate on the carbonyl

Formula 5.2.

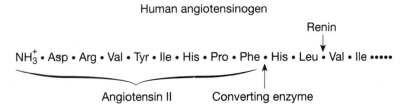

Human angiotensinogen

NH_3^+ • Asp • Arg • Val • Tyr • Ile • His • Pro • Phe • His • Leu • Val • Ile •••••

Renin

Angiotensin II   Converting enzyme

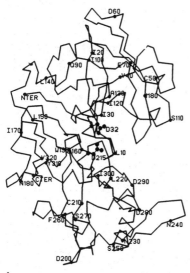
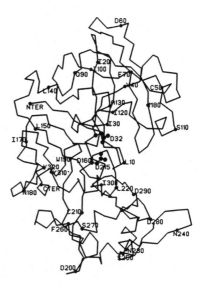

A

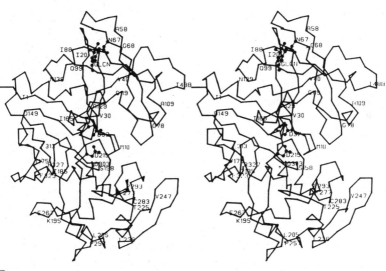

B

Figure 5.6. Stereoview of (A) pepsin and (B) renin, showing their similar tertiary structures. The aspartate sidechains in the active sites are marked by black circles. Black circles also mark every tenth residue along the pepsin chain. In (B) black circles also mark a carbohydrate moiety attached to asparagine 67. (Courtesy of Dr. Anita Sielecki and Professor M. N. G. James)

side and tryptophan, tyrosine, isoleucine, or phenylalanine on the imino side of the scissile bond. Renin, on the other hand, is highly specific for the key leucyl–valyl bond and its surrounds in angiotensinogen. Tantalizingly, the new structure does not yet reveal the stereochemical basis of this high specificity because large thermal motions of the substrate-binding loops obscure the precise positions of important residues relevant for the design of inhibitors with antihypertensive potential.

The difference between the pH profiles of the two enzymes is also very large. Proteolysis by pepsin works fastest between pH2 and 4, while that by renin is fastest between pH 5.5 and 7.5. That difference is not attributable to any specific amino acid replacement in or near the active site, but rather to the different ratios between the total numbers of opposite charges: at neutral pH pepsin carries 43 negative and 5 positive charges, compared to 33 negative and 31 positive ones in renin. These charges are mainly on the enzymes' surfaces. It seems that the electric field generated by the overwhelmingly negative charge of pepsin lowers the pK_as of the aspartates in the active site cleft to optimize catalysis at the stomach's low pH, while in renin that external field is weak.

Several synthetic acid proteinase inhibitors contain the naturally occurring inhibitor statin at the center of an oligopeptide (Formula 5.3). Its incorporation into the 6–13 octapeptide of angiotensinogen yielded some powerful inhibitors of renin, probably because its CHOH mimics the tetrahedral intermediate of natural substrates. This led to synthesis of oligopeptides that simply incorporate an hydroxyethylene -CHOH-CH$_2$- in place of -CO-NH- between leucine 10 and valine 11 of the natural decapeptide substrate. This

Formula 5.3.

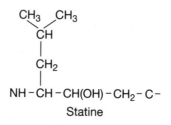

$$\begin{array}{c} CH_3 \quad CH_3 \\ \diagdown \!\! \diagup \\ CH \\ | \\ CH_2 \\ | \\ NH-CH-CH(OH)-CH_2-C- \end{array}$$
Statine

$$NH_3^+ \cdot His \cdot Pro \cdot Phe \cdot His \cdot Leu \overset{H}{\underset{OH}{-C-}} CH_2 \cdot Val \cdot Ile \cdot His \cdot COO^-$$

compound inhibits renin with high specificity and an $I_{50} = 7 \times 10^{-10}$; it was the starting point for further development based on then known structures of acid proteases (Blundell, et al., 1989).

Design of Inhibitors of the Angiotensin-Converting Enzyme
The angiotensin-converting enzyme is an exodipeptidase of 140 Kd containing a single polypeptide chain and one zinc ion. Its catalytic properties are similar to those of two other zinc proteinases: pancreatic carboxypeptidase A and thermolysin. Cushman and his colleagues therefore designed inhibitors on the assumption that carboxypeptidase A and the angiotensin-converting enzyme have similar active sites (Cushman et al., 1977). In the structure of carboxypeptidase A, solved by W. N. Lipscomb and others at Harvard, the zinc ion lies in a pocket where it is coordinated to two histidines, one glutamate and one water molecule. That pocket leads to a tunnel, with an arginine and a cavity for an aromatic sidechain at its head; the arginine binds the C-terminal carboxylate of the substrate (Fig. 5.7) (Lipscomb, 1970). The carbonyl oxygen of the substrate's scissile bond is believed to be polarized by the positive charges of the zinc ion and of another nearby arginine (127), thus facilitating a nucleophilic attack on the carbonyl carbon by a water molecule that is rendered reactive by its two hydrogen bonds to a glutamate. The attack leads to a tetrahedral intermediate and to the donation of one of the water molecule's protons to the leaving amino group (Matthews, 1988) (Fig. 5.8).

The angiotensin-converting enzyme splits off a C-terminal dipeptide in place of the single C-terminal residue cleaved off by carboxypeptidase A. Cushman therefore predicted that its tunnel must be 3.4Å longer, and that it should also contain a hydrogen bond donor and an acceptor to bind to the peptide preceding the substrate's terminus. The first inhibitor designed to fit such an active site was succinylproline, which had an $I_{50} = 3.3 \times 10^{-4}$. Succinylproline bound more strongly than other succinyl amino acids because its pyrrolidine ring restricts rotation about the $C\alpha$–N bond and therefore reduces the loss of rotational entropy on binding. Cushman's group argued that sulphur has a higher affinity for zinc ion than oxygen. They therefore replaced the succinyl carboxyl by a sulphydryl, which led them to the synthesis of 3-mercapto-2-methylpropanoyl-L-prolin (Captopril). This has an I_{50} of 2×10^{-7}M and is widely used as an antihypertensive drug (Cushman et al., 1977) (Fig. 5.9, on page 154).

Captopril has some undesirable side effects attributable to its sulphydryl group. Patchett and others found that the cysteinyl

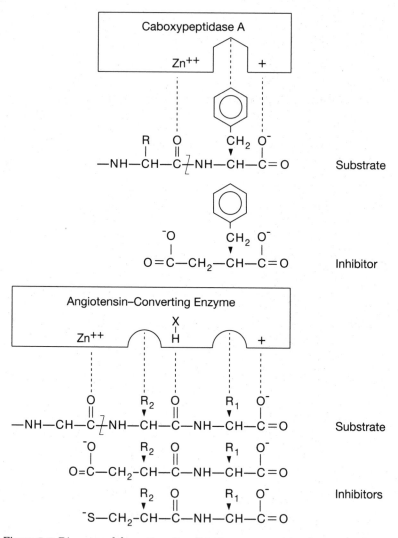

Figure 5.7. Diagram of the active site of bovine pancreatic carboxypeptidase A and the analogous, longer hypothetical one of the angiotensin-converting enzyme. R_1 and R_2 represent sidechains. (Reproduced, by permission, from Cushman et al., 1977)

sidechain can be replaced by a homophenylalanyl moiety without loss of activity, and they synthesized a series of inhibitors based on this concept (Patchett et al., 1980). Monzingo and Matthews crystallized some of these inhibitors bound to the bacterial

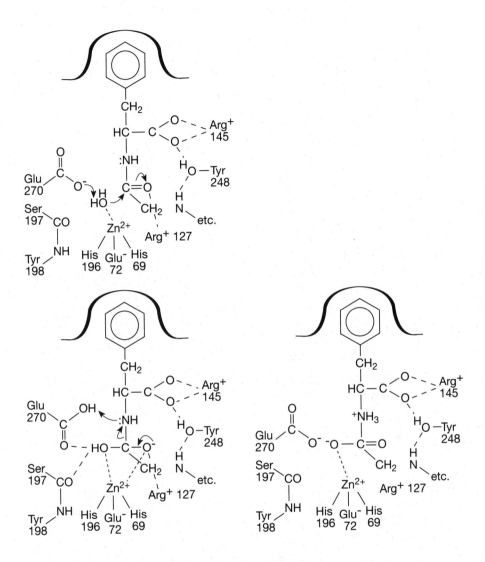

Figure 5.8. Proposed catalytic mechanism of bovine pancreatic carboxypeptidase A. For description see text. (Reproduced, by permission, from Christianson et al., 1987)

endopeptidase thermolysin. X-ray analysis showed the carboxylate of their phenylalanine to be bound to the zinc ion and their NH to donate a hydrogen bond to one of the main chain carbonyls (Monzinga and Matthews, 1984) (Fig. 5.10 on page 155).

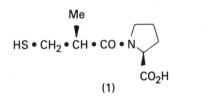

(1)

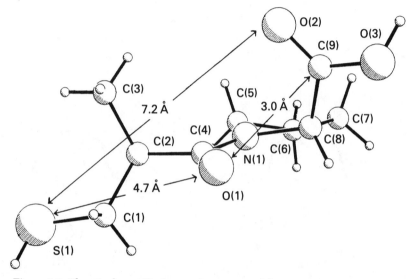

Figure 5.9. Chemical constitution and structure of the angiotensin-converting enzyme inhibitor Captopril. (Reproduced, by permission, from Hassall et al., 1984)

Hassall and others later combined the entropic and the homo-phenylalanyl approaches to design a new and even more potent inhibitor. They fitted a model of Captopril to the accurately known active site of thermolysin and determined the most probably angles taken up by its two freely rotating bonds. They then designed a bicyclic derivative that abolished that freedom of rotation and imposed a conformation exactly mimicking those angles (Fig. 5.11). Finally, they replaced the cysteyl sidechain by a homophenylalanyl sidechain. That synthesis led to the powerful new drug Cilazapril. Hassall's rational approach, based on a known enzyme structure, allowed his group to synthesize and screen fewer than 100 compounds, rather than the more usual

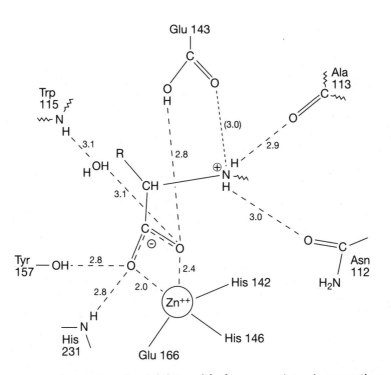

Figure 5.10. Interaction of an inhibitor of the human angiotensin-converting enzyme with the active site of the bacterial zinc endopeptidase thermolysin. The inhibitor is N-(1-carboxy-3-phenylpropyl)-L-leucyl-L-tryptophan. Note the hydrogen bonds between the carboxylate of the inhibitor and the glutamate, tyrosine, and histidine, and its interactions with the zinc ion. (Reproduced, by permission, from Monzinga and Matthews, 1984)

10,000, before they arrived at a clinically useful drug (Hassall et al., 1982; Attwood et al., 1986).

Inhibitors of Leukocyte Elastase

Excessive secretion of human leukocyte elastase or reduced levels of its natural inhibitor α_1-antitrypsin can lead to pulmonary emphysema, rheumatoid arthritis, and other diseases. Elastase also occurs in lysosomes and probably plays an important part in certain inflammations. Its structure has been solved (Bode et al., 1986). As of 1991, several drug companies are trying to design an elastase inhibitor that does not significantly inhibit other physiologically important serine proteinases (Hassall et al., 1985).

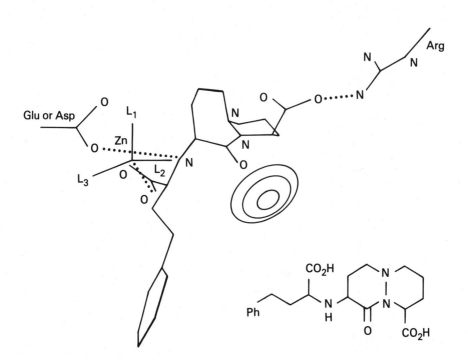

Figure 5.11. Chemical formula and orientation of the antihypertensive drug Cilalapril when bound to the hypothetical active site of the angiotensin-converting enzyme. (Reproduced, by permission, from Attwood et al., 1986)

Another Approach to Drugs Against HIV: Reverse Transcriptase of the Human Immunodeficiency Virus—Structure of Its Ribonuclease H Domain

In 1963 Howard Temin found that the antibiotic actinomycin D, which was known to inhibit the transcription of DNA into RNA, suppressed the growth of Rouse sarcoma virus *in vitro,* even though its genome is made of RNA. Temin concluded that this RNA must be replicated via a DNA intermediate, which then serves as a template for the synthesis of the viral RNA (Temin, 1963). Seven years later both he and Mizutani and, independently, Baltimore confirmed that hypothesis (Temin and Mizutani, 1970; Baltimore, 1970). Their discovery of an enzyme that catalyzes the transcription of RNA into DNA provoked an indignant editorial in *Nature* ("Central Dogma Reversed," 1970) claiming that molecular biologists had gotten it wrong:

Central Dogma Reversed

The central dogma, enunciated by Crick in 1958 and the keystone of molecular biology ever since, is likely to prove a considerable over-simplification. That is the heretical but inescapable conclusion stemming from experiments done in the past few months in two laboratories in the United States. For the past 20 years the cardinal tenet of molecular biology has been that the flow or transcription of genetic information from DNA to messenger RNA and then its translation to protein is strictly one way. But on pages 1209 and 1211 of this issue of *Nature*, Baltimore and Mizutani and Temin claim independently that RNA tumour viruses contain an enzyme which uses the viral RNA as a template for the synthesis of DNA and thus reverses the direction of genetic transcription.

RNA transcriptase performs three closely linked functions: it catalyzes the RNA-directed synthesis of a single strand of DNA with a base sequence complementary to the RNA template; it then hydrolyzes the RNA template and catalyzes the synthesis of a second strand of DNA with a base sequence identical to that of the template, from which messenger RNA is eventually transcribed, thus preserving the central dogma of molecular biology that information flows from DNA to RNA to protein, despite nature's scorn.

The transcriptase of the human immunodeficiency virus is cleaved from a polyprotein by the acid proteinase described earlier in this chapter. The same proteinase then splits off a C-terminal 14kD fragment from a proportion of the transcriptase molecules; the remaining 51kD N-terminal portion forms a heterodimer with the full-size 66 kD protein. The N-terminal part of that heterodimer contains the DNA-polymerase activity, but its amino acid sequence shows no homology with those of other DNA polymerases. The C-terminal domain contains the ribonuclease activity specific for RNA in RNA–DNA–hybrid form (Goff, 1990).

Its amino acid sequence shows homology with the ribonuclease domains of other retroviruses and also with those of the free ribonucleases H of *E. coli* and yeast (Fig. 5.12). First, two independent groups determined the structure of the *E. coli* enzyme which was easier to isolate, and then another group followed with the structure of the ribonuclease H domain of the HIV transcriptase itself (Yang et al., 1990; Katayanagi et al., 1990 Davies et al., 1991). Both structures are made of a five-stranded β-sheet flanked by four α-helices, but they differ by the deletion of 19 residues from external loops of the transcriptase domain (Fig. 5.13).

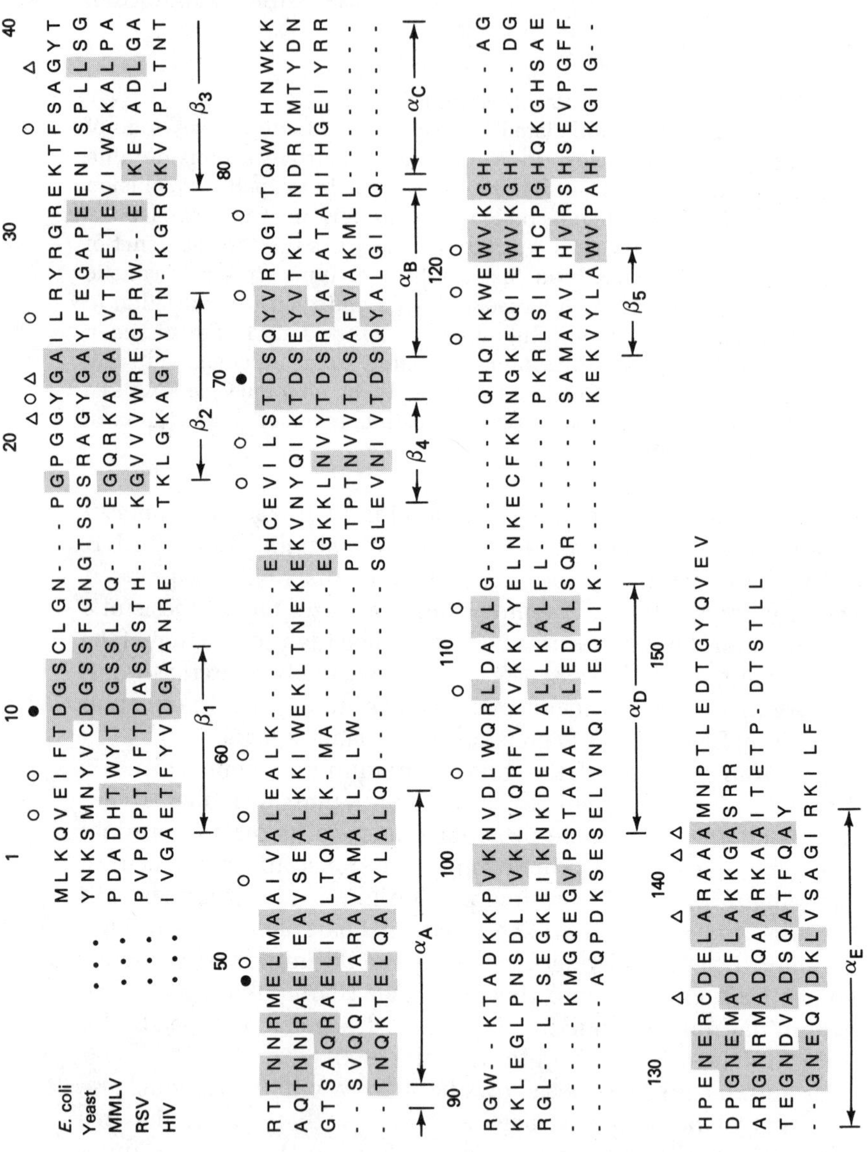

Figure 5.12. Amino acid sequences of ribonucleases H from the bacterium Escherichia coli, *the yeast* Saccharomyces cerevisiae, *and the retroviruses Moloney murine leukemia virus (MMLV), Rous sarcoma virus (RSV), and human immunodeficiency virus (HIV). The sequences have been aligned so as to leave no insertions or deletions within α-helices or β-strands. The shaded areas indicate homologies. The circles mark residues in the hydrophobic cores. The asterisks mark the aspartates and the glutamate in the active site. The triangles mark contacts between helix E and the pleated sheet. (Reproduced, by permission, from Yang et al., 1990)*

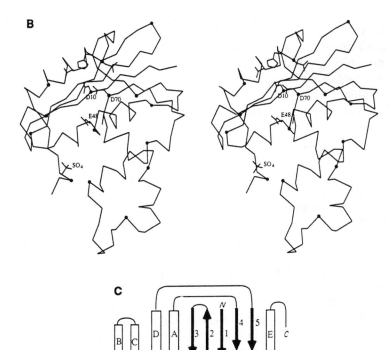

Figure 5.13. Course of the polypeptide chain in ribonuclease H of Escherichia coli *shown diagrammatically in (A) and in stereo in (B). The black circles mark every tenth α-carbon. The positions of the two aspartates and the glutamate in the active site are labelled and their sidechains are drawn, as are those of valine 121, histidine 124, and aparagine 130. (C) shows the way the helices and β-strands are connected and the notation used for ribonucleases H. (Reproduced, by permission, from Yang et al., 1990)*

The active site lies in a shallow groove that contains two divalent metal ions coordinated to three aspartates and a glutamate; further from the metals lie a histidine and a serine (Fig. 5.14). Replacement of either of the aspartates 443 or 498 or of the glutamate by neutral residues inhibits catalysis, but replacements of the histidine, or the third aspartate, merely slow it down.

The catalytic mechanism has been inferred indirectly. The active site of ribonuclease H is similar to that of a fragment split from DNA polymerase I of *E. coli;* that fragment cleaves single nucleotides from the 5'-ends of single-stranded DNA by breaking the bond between the 3'-oxygen of the leaving ribose sugar and the phosphorus attached to the 5'-oxygen of the next ribose sugar along the chain. It is known as the Klenow fragment.

Beese and Steitz have solved the structures of the Klenow fragment alone and those of its complexes with a deoxythymidine-tetranucleotide, representing its substrate, and deoxythymidine-monophosphate (dTMP), representing the leaving group after catalysis. These complexes were stable enough to be crystallized because Beese and Steitz used an inactive mutant enzyme that had one of its catalytic aspartates replaced by an alanine, but had a structure identical to that of the native enzyme (Beese and Steitz, 1991).

The metals in the active site of ribonuclease H were first thought to be magnesium ions, but it must have seemed improbable to Beese and Steitz that this was true of the Klenow fragment, because we now know of about 80 enzymes that rely on zinc ions for catalysis, while magnesium ions tend to play a structural rather than a catalytic role in the active sites of enzymes. The reason is this. The eight electrons in the 2s and 2p orbitals of the magnesium ion effectively screen its nuclear charge; its polarizing power is therefore much weaker than that of the zinc ion, which contains 18 more protons but has an ionic radius of 0.1Å *smaller* than that of the magnesium ion. It attracts electrons back from its anionic ligands, which gives the formally ionic metal–ligand bond a partly covalent character and makes the metal into an acid; metal ions with that property are called, after its discoverer G. N. Lewis, *Lewis acids.* For example, the magnesium hydrate, Mg^{2+}aq, has a pK_a of 11.5, whereas zinc hydrate, Zn^{2+}aq, has a pK_a of 8.5, so, it will dissociate into $ZnOH^+ + H^+$ a thousand times more easily on titration with alkali. There is another difference between the two metal ions. Magnesium is normally coordinated to six atoms at the corners of an octahedron, while zinc is coordinated to four atoms at the corners of a tetrahedron.

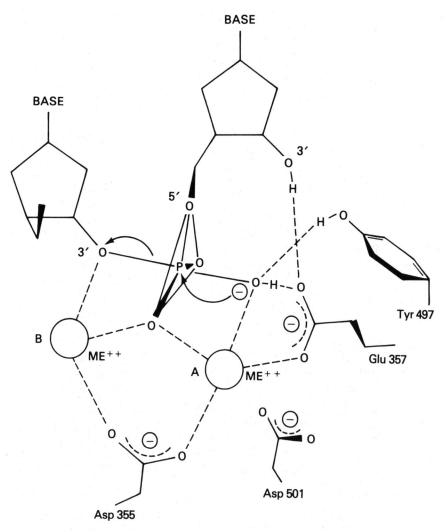

Figure 5.14. Diagram of the active site of the Klenow fragment of polymerase I of E. coli. *Zinc is bound to site A and magnesium to site B. For explanation see text. (Reproduced, by permission, from Beese and Steitz, 1991)*

Sometimes this tetrahedron is irregular to allow a fifth atom to squeeze in between. Zinc therefore polarizes it surrounding atoms more strongly, which makes it a better catalyst.

Beese and Steitz soaked a metal-free crystal of the Klenow fragment plus dTMP in a mixture of $MgCl_2$ and $ZnCl_2$. The resulting electron density map showed the heavier zinc ion to be tetrahedrally coordinated to two of the aspartates, a water molecule and one of the phosphate oxygens, while the lighter magnesium ion was octahedrally coordinated to three aspartates, one glutamate and two of the phosphate oxygens.

This structure suggested a plausible catalytic mechanism. In figure 5.14 site B would be occupied by magnesium and site A by zinc; it has been given a negative sign to indicate its function as a Lewis acid. Its poorly screened positive nuclear charge is thought to expel a proton from the bound water molecule and turn it into a hydroxyl ion. This hydroxyl is hydrogen-bonded to glutamate 357 and points its lone pair electrons at the phosphorus atom of the substrate. This would favor a nucleophilic attack on the phosphorus and would generate a transition state with the phosphorus coordinated to five oxygens at the corners of a trigonal bipyramid, indicated in the figure by the triangle and by the straight lines to the 3′ oxygen and the OH^-. The transition state would disintegrate with rupture of the 3′O–P bond, leaving a negative charge on 3′O which is stabilized by the close-by magnesium ion.

Even though the Klenow fragment splits DNA rather than RNA, the homology of the HIV ribonuclease's active site to that of the Klenow fragment implies that the two divalent metal ions should be the same in both proteins, one site being occupied by a magnesium ion and the other by a zinc ion, and that their catalytic mechanisms should also be similar, with histidine 639 in the ribonuclease H taking the place of tyrosine 497 in the Klenow fragment. That histidine must fulfill a function, perhaps as a proton donor, because its replacement by other residues impairs the function of the ribonuclease H.

Thanks to elaborate proofreading mechanisms, the error rates in the replication of DNA are as low as one mismatched base pair in 10^9, but reverse transcriptases lack such mechanisms and copy sequences with error rates as high as 1 in 10^3 or 10^4, no matter whether they copy RNA or DNA. This lack of fidelity results in a very high mutation rate of the human immunodeficiency virus which complicates the development of effective drugs and vaccines.

The detailed understanding of the structure of the Klenow fragment combined with its substrate or its leaving group, and of its likely catalytic mechanism complements the information derived from the structure of the HIV ribonuclease H domain. By the end of 1991 Thomas A. Steitz and his colleagues at Yale University had solved the structure of the complete HIV transcriptase molecule, which should allow new antiviral drugs to be made to measure.

Further Reading

Arnold, E. and G. F. Arnold, 1991. Human immuno-deficiency virus structure: Implication for antiviral design. *Adv. Virus Research* **39**:1–88.

Hol, W. L. J. 1986. Protein crystallography and computer graphics: Toward rational drug design. *Angewandte Chemie* **25**:767–778.

Koch, M. G. 1987. *AIDS: Vom Molekül zur Pandemie.* Spektrum der Wissenschaft, Heidelberg.

6

How Mutations Can Impair Protein Function: Molecular Pathology of Human Hemoglobin and Other Proteins

Chemists often ask whether the structure of proteins in the crystal is the same as in solution. Protein crystals are like sponges. X-ray analysis has shown that in most crystals protein molecules make narrow contacts with each other while most of their surfaces are covered with water. The lattice energies of protein crystals are small compared with the energies that stabilize the protein molecules, too small to disturb their internal structures significantly although their surface features may be affected.

This is a theoretical argument. Its first experimental confirmation came from the study of the abnormal human hemoglobins. Most of these arise from single nucleotide base substitutions in the globin genes that cause substitutions of single amino acid residues in one or the other pair of globin chains. H. Lehmann and I found an exact correlation between the changed physiological properties of the abnormal hemoglobins on the one hand and the stereochemical perturbations found by X-ray crystallography on the other hand. For example, replacements of external amino acid residues were usually harmless, but interior replacements produced symptoms. Such a correlation could not have existed if the structure of hemoglobin in the crystal had differed significantly from that in the red blood cell; it was also encouraging that many clinical symptoms could be interpreted at the atomic level (Perutz and Lehmann, 1968; Morimoto et al., 1971).

The two most frequent hemoglobin diseases are thalassemia and sickle cell anemia. Thalassemia is generally caused by deficient synthesis of either α- or β-globin, rather than by an altered structure of hemoglobin. Sickle cell anemia is caused by the substitution of glutamate 6β by valine. The substitution is external and causes no

significant alterations of the internal structure of hemoglobin, but
it causes the deoxygenated hemoglobin to polymerize in the red
cell and to precipitate in the form of long fibers. At high hemoglobin
concentrations and either very low or very high ionic strength,
normal human deoxyhemoglobin also has a tendency to polymerize into long chains; these chains then aggregate to form
perfect three-dimensional crystals. Filaments of sickle cell hemoglobin contain the same long chains as the crystals of normal
hemoglobin, but the chains are paired and seven of these paired
chains aggregate to form long fibers. The pairs are held together
by a hydrophobic plug formed by the sidechain of valine 6β of one
member of each pair that fits into a hydrophobic socket in the
opposite member of the pair. This unique hydrophobic contact
appears to be the sole cause of the low solubility of sickle cell
hemoglobin (Fig. 6.1) (Schechter et al., 1987).

*Figure 6.1. Structure of double filament of hemoglobin S. (A) General
arrangement. (B) Contact of Val 6β of one molecule with Phe 85β and
Leu 88β of its neighbor. (C) Position of glutamate sidechain of 6β in
normal human hemoglobin A. It does not combine with that neighboring
hydrophobic site. (Reproduced, by permission, from Schechter et al., 1987)*

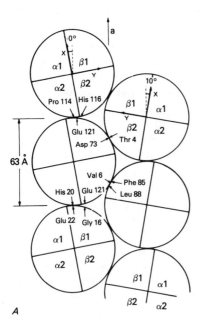

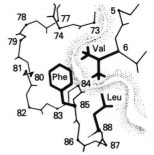

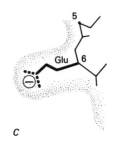

Directed mutagenesis has made it possible to test this interpretation. Baudin-Chich and others have replaced glutamate 6β by isoleucine which differs from valine by only one additional methyl group. The solubility of this hemoglobin in 1.8M phosphate is half that of sickle cell hemoglobin (Baudin-Chich et al., 1990).

When a hydrophobic amino acid residue that is external in the free hemoglobin molecule becomes buried on polymerization, the free energy lost should be similar to that lost on transferring that same amino acid from water to a hydrophobic solvent. The observed difference in the free energy of transfer between valine and isoleucine is $\Delta\Delta G$ = 800 cal/mol. If the ratios of the solubilities S of the two deoxyhemoglobins reflected the difference in free energy of transfer from water to a nonpolar solvent at 20°C, then $\log (S_{Val}/S_{Ile}) = \dfrac{\Delta\Delta G}{2.30RT} = \dfrac{\Delta\Delta G}{1340}$ if $\Delta\Delta G$ is given in calories. Substituting the observed $\Delta\Delta G$ of transfer gives S_{Val}/S_{Ile} = 4, which suggests that the burying of the extra methyl group is sufficient to explain the halved solubility of the isoleucine mutant. The example shows that the solubility of proteins is more likely to be a function of the degree of complementarity of the surfaces of neighboring molecules rather than simply a function of the ratio of hydrophilic to hydrophobic residues. Knowledge of the detailed structure of sickle cell hemoglobin has not yet led to the design of an antisickling drug that is effective and sufficiently nontoxic to be administered to patients in the large doses that would be needed to achieve a significant effect, because the plasma concentration of hemoglobin is about 2mM (tetramer), whereas most drugs become toxic at plasma concentrations above 10nM.

Many hemoglobin diseases arise from disturbances of the allosteric equilibrium between its two alternative structures: the deoxy- or T-structure with a low oxygen affinity, and the oxy- or R-structure with a high oxygen affinity. Mutations that break bonds that stabilize specifically the oxy- or R-structure reduce the oxygen affinity; conversely, mutations that break bonds in the deoxy or T-structure raise the oxygen affinity. The concentration of red cells in the blood is regulated by a sensor in the kidneys that responds to the oxygen tension. If the oxygen affinity of hemoglobin is too high, and not enough oxygen is released, then the sensor causes more of the hormone erythropoietin to be released, and that in turn stimulates red cell production. In consequence, high oxygen affinity causes polycythemia, or too high a concentration of red cells. Conversely, abnormal hemoglobins with low oxygen affinity can cause anemia.

Figure 7.8 on page 182 illustrates two abnormal hemoglobins which disturb the switch between the R- and T-states in opposite directions. In the R-structure the switch is held in place by a hydrogen bond between aspartate 94α and asparagine 102β. Replacement of the asparagine by threonine destabilizes the R-structure and therefore causes low oxygen affinity and anemia. In the T-structure the switch is held in place by a hydrogen bond between tyrosine 42α and aspartate 99β. Replacement of the aspartate by asparagine causes high oxygen affinity and polycythemia. The rise in oxygen affinity allows one to calculate the free energy contribution of the hydrogen bond to the T-structure. It is about 3 kcal/mol (Fermi and Perutz, 1981). Table 6-1 shows other substitutions that raise or lower the oxygen affinity. Many errors of metabolism may be due to mutations that disturb the allosteric equilibria of other vital proteins in a similar way.

Many abnormal hemoglobins are unstable. They denature and precipitate in the red cell, shortening its life span and causing hemolytic anemia of various degrees of severity. Table 6.2 lists the

Table 6.1A. Causes of High Oxygen Affinities

Cause	Number of different mutants observed
Loss of hydrogen bonds, salt bridges, and nonpolar interactions that stabilize the structure as a whole	18
Loss of bonds that stabilize only the T-structure	20
General disruption of structure	11
Loss of DPG affinity	7
Influence on position or tilt of heme and replacements of distal residues and related effects	6
No explanation	5
Electrostatic effects on allosteric equilibrium	2
Loss of heme	1
Total	70

6.1B. Causes of Low Oxygen Affinities

Cause	Number of different mutants observed
Loss of bonds that stabilize only the R-structure or gain of extra bonds that stabilize the T-structure	5
Influence on position or tilt of heme or on position of distal residues, including allosteric effects caused by these	14
Electrostatic effects on allosteric equilibrium	5
No explanation	4
Total	28

causes of such instability. Again, similar mutations must be the causes of malfunctions in many other human proteins, but few have so far been documented in precise stereochemical terms. One of them is α_1-antiproteinase, whose pathology has been discussed in Chapter 2 on page 70. Another is insulin, where mutants indicated which face of the molecule binds to the insulin receptor (Baker et al., 1988).

The most common form I of collagen contains two chemically distinct polypeptide chains α_1I and α_2I. Collagen fibers in bone, tendon, skin, and cornea consist of molecular cables formed by three polypeptide chains, $2\alpha_1$I + $1\alpha_2$I, coiled around each other. Their amino acid sequences are dominated by the repeating motif glycine-X-Y, where X is often proline and Y either proline or hydroxyproline. Each chain has a structure that is a twisted version of the polyproline helix shown in figure A2.5, page 286. The three tightly packed chains are held together by van der Waals interactions and by hydrogen bonds between the NH and CO groups of glycines in neighboring chains. The glycines are packed so tightly that no other residue fits in their place (Rich and Crick, 1955). Mutations that cause one of the glycines to be replaced by another residue give rise to brittle bone disease, even in heterozygotes, because a single such substitution in even one of the three chains retards their assembly and thereby causes chemical and structural changes of the collagen fibers (Engel and Procktop, 1991).

A similar effect was found in heterozygotes for a hemoglobin mutation that caused glycine B6(24)β to be replaced by a valine.

Table 6.2. Causes of Instability in Abnormal Hemoglobins

Cause	Number of different mutants observed
Loss of nonpolar "plug" that normally seals the protein surface	16
Loss of hydrogen bonds or salt bridges	7
Introduction of interior charge or dipole	5
Introduction of interior gap	6
Introduction of wedge between helices	6
Misfit at subunit contact	4
Introduction of Pro into α-helix	13
Unclear	2
Introduction of side chain at position of invariant Gly in β-bend	1
Deletions	11
Other causes	2
Total	73

This glycine packs tightly against another glycine, E8(64)β. Close packing of helices B and E at this point is essential for the stability of the globin structure. Substitutions of bulkier residues that prise the helices apart cause denatured hemoglobin to accumulate in the red cell and to form inclusion bodies. They lead to premature destruction of the red cell and hemolytic anemia (Huisman et al., 1971). In this way genetically recessive mutations can become dominant in their clinical manifestations.

Many amino acid replacements and deletions that impair function have been mapped in the sequences of other variant human proteins. For example, Giannelli et al. (1991) list 177 different variants of human clotting factor IX, and Vulliami, Mason, and Luzzatto (1992) list 32 in human glucose-6-phosphate dehydrogenase. The mapping of the human genome will greatly expand this catalogue, but in the absence of three-dimensional structures the malfunctions caused by such mutations will not be understood in stereochemical terms.

Further Reading

Bunn, H. F. and B. G. Forget 1986. *Hemoglobin: Molecular, Genetic and Clinical Aspects*. W. B. Saunders Co., Philadelphia.

Eaton, W. A. and J. Hofrichter 1990. Sickle cell polymerization. *Adv. Protein Chem.* **40**:63–249.

Engel, J. and D. J. Prockop 1991. The zipper-like folding of collagen triple helices and the effects of mutations that disrupt the zipper. *Ann. Rev. Biophys. Chem.* **20**:137–52.

Gianelli, F., P. M. Green, K. A. High, S. Sommer, D. P. Lillicrap, M. Ludwig, K. Oleg, P. H. Reitsma, M. Gossens, A. Yoshioka and G. G. Brownlee 1991. Haemophilia B: Database of point mutations and short additions and deletions, 2nd edition. *Nucl. Acids Res.* **19**:2193–2219.

Huisman, T. H. J., A. K. Brown, G. D. Efremor, J. B. Wilson, C. A. Reynolds, R. Uy and L. L. Smith 1971. Hemoglobin Savannah (B6(24)β Glycine→ Valine): An unstable variant causing anemia with inclusion bodies. *J. Clin. Invest.* **50**:650–59.

Perutz, M. F. 1990. Mechanisms regulating the reactions of human hemoglobin with oxygen and carbon monoxide. *Ann. Rev. Physiol.* **52**:1–25.

Rich, A. and F. H. C. Crick 1955. The structure of collagen. *Nature* **176**:915–16.

Stamatoyannopoulos, G., A. W. Nienhuis, P. Leder, and P. W. Majerus 1987. *The Molecular Basis of Blood Diseases*. W. B. Saunders, Philadelphia.

Vulliamy, T., P. Mason and L. Luzzatto 1992. The molecular basis of glucose 6-phosphate dehydrogenase deficiency. *Trends in Genetics* **8**:138–43.

7

How Gene Technology Can Improve Protein Function

Engineering a Better Insulin

In 1935 Dorothy Hodgkin, having moved from Cambridge to Oxford, crystallized insulin and put one of her small crystals in front of an X-ray beam. That night, when she developed the film, she saw minute, regularly arranged spots forming a diffraction pattern that held out the prospect of solving insulin's structure. Later she wandered around the deserted streets, madly excited that she might be the first to determine the structure of a protein, but next morning she woke up with a start: could she be sure that her crystals really were insulin rather than some trivial salt? She rushed back to the lab before breakfast. A simple spot test on a microscope slide showed that her crystals took up the ninhydrin stain characteristic for protein, proving her fears to have been groundless.

As insulin was a smaller protein than pepsin, she hoped that it would be easier to solve, but she was mistaken. It took her 34 years (Blundell, et al., 1971; Baker et al., 1988). Insulin consists of two polypeptide chains, one containing 21 and the other 30 amino acid residues. The two chains are linked by a disulphide bridge (Fig. 7.1). Figures 7.2 and 7.3 show the tertiary structure; the A-chain forms two α-helices connected by a short piece of straight chain; the B-chain starts with two short stretches of straight chain, followed by an α-helix and a long β-strand. *In vivo* the monomer is the active form, but in solution and in the crystal two monomers coalesce to form a dimer, and three dimers join to form a hexamer. In the dimer, the C-terminal strands of two B-chains are joined by hydrogen bonds to form an antiparallel pleated sheet.

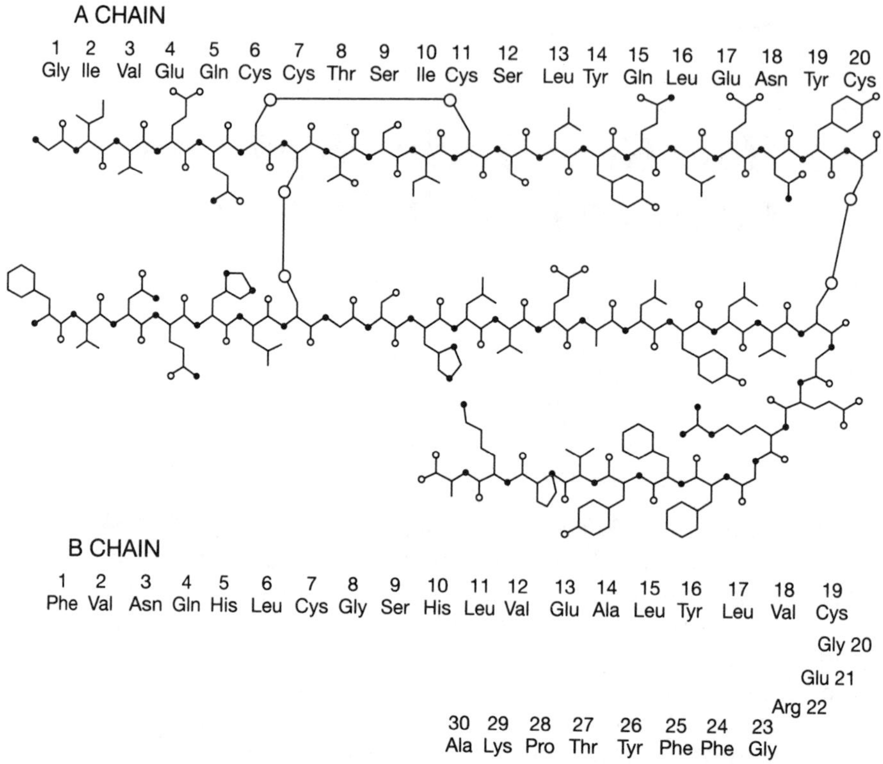

A CHAIN

1	2	3	4	5	6	7	8	9	10	11	12	13	14	15	16	17	18	19	20
Gly	Ile	Val	Glu	Gln	Cys	Cys	Thr	Ser	Ile	Cys	Ser	Leu	Tyr	Gln	Leu	Glu	Asn	Tyr	Cys

B CHAIN

1	2	3	4	5	6	7	8	9	10	11	12	13	14	15	16	17	18	19
Phe	Val	Asn	Gln	His	Leu	Cys	Gly	Ser	His	Leu	Val	Glu	Ala	Leu	Tyr	Leu	Val	Cys

Gly 20

Glu 21

Arg 22

30	29	28	27	26	25	24	23
Ala	Lys	Pro	Thr	Tyr	Phe	Phe	Gly

Figure 7.1. Chemical constitution of pig insulin used for determination of the crystal structure. In human insulin alanine B30 is replaced by threonine. (Reproduced, by permission, from Baker et al., 1988)

The insulin receptor is a glycosylated protein of 350–400 kilodaltons that spans cell membranes and is made up of two pairs of chains: the α-chains with 719 and the β-chains with 619 amino acid residues each. At the center of the β-chains is a sequence of 23 mainly hydrophobic residues that signifies a membrane-spanning α-helix. The α-chains appear to be extracellular, while the β-chains contain one extracellular and one cytoplasmic domain. The latter is a tyrosine kinase (Fig. 7.4) (Ulrich et al., 1985). Affinity cross-linking and photo-affinity labelling have shown insulin to bind to the α-chains. That binding activates the kinase of the β-chains, which in its turn initiates a cascade of phophorylation, so that the activation of a single insulin receptor molecule may activate many other proteins and these in turn may

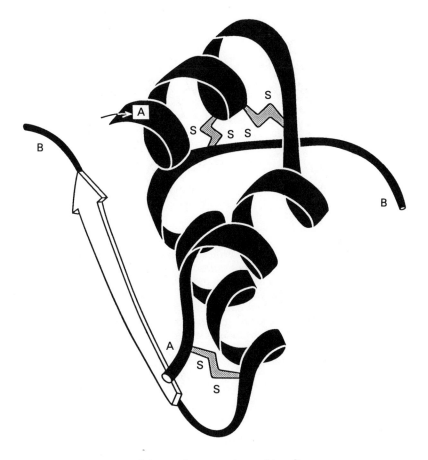

Figure 7.2. Diagram of the tertiary structure of insulin.

activate the transcription of genes. The insulin receptor must be an allosteric protein, i.e., one that is in equilibrium between active and inactive structures. By biasing that equilibrium toward the active structure, insulin behaves as an allosteric effector, like 2,3-diphospho-glycerate in hemoglobin or ATP in aspartate transcarbamylase (Perutz, 1989). Electron micrographs show the receptor to be T-shaped, but after combination with insulin the bar of the T seems to kink in the middle to make the receptor Y-shaped. The bar probably represents the α-subunits (Christianson et al., 1991).

The amino acid sequence of the tyrosine kinase domain of the insulin receptor shows strong homology with that of the epidermal

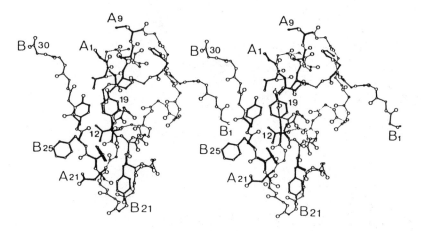

Figure 7.3. Stereoview of insulin from the same direction as in Figure 7.2, showing the residues that are important in insulin's binding and action (bold lines). For the rest of the molecule only the main chain is shown, the A chain as thin lines and the B chain as thick lines. In the insulin dimer the β-strands from B22 to B25 of neighboring monomers are joined by hydrogen bonds, forming an antiparallel pleated sheet. Replacement of phenylalanine B25 by leucine reduces the activity 100-fold; replacement of valine B12 by isoleucine reduces it ten-fold. The positions of glycine A1, serine A9, tyrosine A19, asparagine A21, phenylalanine B1, glutamate B21, phenylalanine B25, and alanine B30 are marked. (Reproduced, by permission, from Baker et al., 1988)

growth factor receptor and with the tyrosine kinases that are the products of the oncogenes of the Rous sarcoma (src) family of tumor viruses.

The structure of insulin alone does not reveal the nature of its interaction with its receptor, but the varying degrees of activity of natural and engineered mutants have allowed investigators to map the amino acid sidechains that touch the receptor, or whose replacement alters the conformation of insulin in a way that affects its affinity for the receptor (Fig. 7.3). The face of the molecule that binds to the receptor is the same as that which forms the contact between the dimers, i.e., the left-hand face of Figures 7.2 and 7.3. This is the reason why dissociation into monomers is needed for activity.

It seemed important, therefore, to find out if the structure of insulin monomers in solution is the same as in the hexamers in the crystal. A nuclear magnetic resonance study of a 1mM solution of human insulin in acetic acid of pH 1.5 showed this to be true within

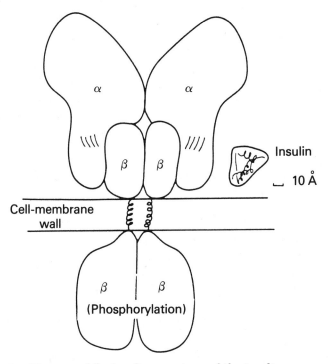

Figure 7.4. Diagram of the insulin receptor and the insulin monomer drawn approximately to scale. (Reproduced, by permission, from Baker et al., 1988)

experimental error. The next question was whether this structure is maintained also on binding to the receptor. A variety of observations suggests that this holds for the bulk of the molecule, but not for the C-terminal β-strand of the B-chain (Fig. 7.3). For example, insulin from which the C-terminal five residues have been cleaved remains fully active. Insulin in which phenylalanine B24 has been replaced by glycine retains three quarters of its activity, even though the entire β-strand from B10 to B30 becomes disordered in solution. On the other hand, an engineered insulin that has lysine B29 covalently linked to glycine A1 by a glycine bridge is inactive, even though that link leaves its structure practically undisturbed. Between them, these observations suggest that biological activity requires the C-terminal β-strand of the B-chain to be mobile, perhaps in order to bring the surface underneath into contact with the receptor (Derewenda et al., 1991; Hua et al., 1991).

When insulin is injected subcutaneously into a diabetic patient, the glucose level in the serum declines slowly over two hours, because the hexamer and higher aggregates are apparently slow to dissociate into monomers. By contrast, insulin secreted by the pancreas in healthy individuals lowers that level almost immediately. The Danish pharmaceutical firm Novo, together with G. Dodson and his associates at York, England, have engineered a human insulin with a faster response by introducing acidic amino acid residues at the boundary between adjacent monomers that give rise to electrostatic repulsion between them (for example, by replacing proline B28 by aspartate) (Fig. 7.5). Measurement of the circular dichroism of solutions of native and mutated insulin showed that the electrostatic repulsion due to that replacement raised the dissociation constant of the oligomers into monomers (Fig. 7.6). When the mutated insulin was injected into pigs, the time taken for the plasma insulin level to reach its peak was halved, which made the blood sugar concentration drop much

Figure 7.5. Boundary between two monomers in a genetically engineered insulin in which proline B28 has been replaced by aspartate. Electrostatic repulsion between that aspartate in the monomer on the right and glutamate A4 in the monomer on the left has raised the dissociation constant of the dimer. The boundary between the monomers runs down the middle. (Reproduced, by permission, from Brange et al., 1988)

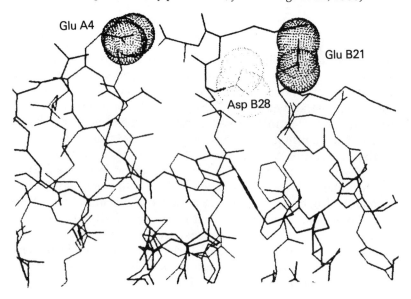

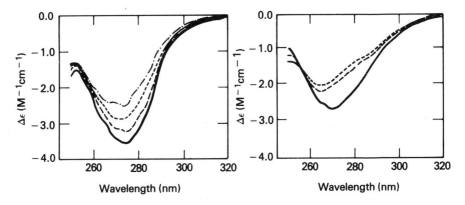

Figure 7.6. Circular diachroism spectra of native (left) and genetically engineered (right) human insulins at concentrations of ---- 3.0×10⁻⁶ M; ----- 3.0×10⁻⁵ M; ---- 3.0×10⁻⁴ M; and —— 3.0×10⁻³ M. The trough at 375 nm arises from interaction between tyrosines in neighboring monomers that disappears with increasing dilution. The shallowness of the trough in the right-hand diagram is evidence of increased dissociation into monomers. (Reproduced, by permission, from Brange et al., 1988)

sooner (Fig. 7.7) (Brange, et al., 1988). This insulin is now in clinical trial, and if successful, should be of great benefit to diabetics. It could not have been designed without knowledge of the three-dimensional structure.

It has recently become apparent that insulin is a member of an entire family of proteins of similar structure but different functions. This includes the insulin-like growth factors IGF I and II, and the hormone relaxin that softens the cervix and the tissue between the two bones at the pubic symphysis in the late stages of pregnancy in preparation for giving birth. Eigenbrot and others (1991) have crystallized human relaxin and solved its structure by X-ray analysis. The structure of the monomer is similar to that of insulin, except that the C-terminal residues of the B-chain are coiled rather than extended. Like insulin, relaxin forms dimers, but their interface is unrelated to that of insulin. In the dimer, the N-terminal tyrosines of neighboring A-chains are bridged by a hydrogen-bonded water molecule, and arginines B9 and 13 of one monomer form hydrogen bonds with the A chains' C-terminal carboxylates of the other monomer. The two arginines are also essential for the interaction of relaxin with its receptor. In insulin, on the other hand, residue B9 is a serine and B13 is a glutamate. There is no evidence that either interacts with the insulin receptor.

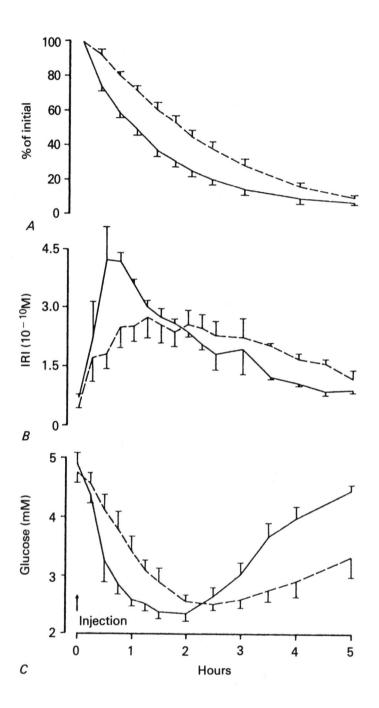

Engineering a Hemoglobin Infusible as a Blood Substitute

The average lifetime of human erythrocytes is only 120 days. In an adult, they break up at the rate of nearly 3 million per second and liberate their hemoglobin into the plasma. The $(\alpha\beta)_2$ tetramer dissociates into $\alpha\beta$ dimers with a dissociation constant, at physiological ionic strength, of $0.7\mu M$ tetramer. The dimers combine noncooperatively with oxygen and exhibit no Bohr effect. This is irrelevant in the erythrocyte where the hemoglobin concentration is over 5mM tetramer, but when red cells break up naturally and their hemoglobin is diluted into the plasma, dissociation takes place at the $\alpha_1\beta_2$ contact pictured in Figure 7.8. The $\alpha\beta$-dimers are scavenged by haptoglobin and eventually broken down. On the other hand, when hemoglobin is transfused into the plasma at concentrations exceeding that of haptoglobin, the $\alpha\beta$-dimers are not broken down; instead, they pass rapidly through the kidneys and are excreted in the urine, because the kidneys are permeable to proteins of molecular weight below 50 kD unless they are very highly charged. This makes the half-life of transfused hemoglobin in the circulation very short and poisons the kidneys.

Further difficulties arise from the high intrinsic oxygen affinity of hemoglobin and the heme irons' susceptibility to autoxidation. In the erythrocyte the allosteric effector 2,3-diphosphoglycerate lowers the oxygen affinity to the physiologically desirable level of p50=26($\pm$1)mmHg. It combines with the deoxy or T-structure in the ratio one mole per mole tetramer, with a dissociation constant of $25\mu M$. In the plasma, it would not be possible to maintain concentrations of 2,3-diphosphoglycerate high enough to saturate transfused hemoglobin. In its absence, the oxygen affinity would rise to a p50 ~ 10mmHg, which is too high for effective delivery of oxygen to the tissues. Before the advent of recombinant gene technology, this property alone made hemoglobin unsuitable for transfusion.

The iron atoms in hemoglobin are ferrous; only ferrous heme combines reversibly with molecular oxygen. In the erythrocyte the

Figure 7.7. (opposite) (A) Absorption rate after subcutaneous injection into n pigs of mutant (——) and normal (-----) human insulin (mean $\pm$ s.e.m., n=8). (B) Plasma insulin after subcutaneous injection of B28Asp mutant (——) and normal human insulin (-----) (13 nmol per kg) into fasted pigs (mean $\pm$ s.e.m., n=6). (C) Plasma glucose after subcutaneous injection of B28Asp mutant (——) and normal human insulin (-----) (6.5 nmol per kg) into fasted pigs (mean $\pm$ s.e.m., n=6). (Reproduced, by permission, from Brange et al., 1988)

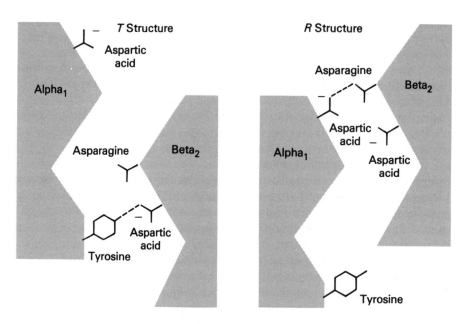

Figure 7.8. Hydrogen bonds between α_1 *and* β_2 *subunits in the T- and R-structures of human hemoglobin.*

heme irons autoxidize at a rate of 2–3% per day and are reduced again by the NADH-dependent enzyme system of methemoglobin reductase. This is absent from plasma, so that autoxidation would proceed unchecked.

Chemistry and gene technology have surmounted the difficulties of fast excretion and high oxygen affinity, but not that of autoxidation. Dissociation into $\alpha\beta$-dimers has been prevented by engineering a variety of covalent links between the subunits, for example, by exploiting the proximity of the C-terminal arginine of one α-chain to the N-terminal valine of its neighbor. Looker et al., (1992) have genetically engineered a glycine linking these two residues so that the two separate α-chains have been replaced by a continuous chain of double the length. This associates with two β-chains to a molecule that is isomorphous with normal hemoglobin A and combines cooperatively with oxygen. X-ray analysis showed that the glycine bridge disturbs the deoxy structure only in its immediate vicinity.

An alternative stratagem consists in the introduction of cysteines placed opposite each other in the $\alpha_1\beta_2$ contact so that the planes of the two cysteyl sidechains are inclined to each other at the correct dihedral angle of 90°. Other disulphide bridges can be introduced to link several tetramers together into an oligomer of higher order, mimicking the invertebrate hemoglobins of high molecular weight that are freely dissolved in the plasma.

The oxygen affinity of these cross-linked hemoglobins can be reduced by mimicking one of the natural human mutants with low oxygen affinity. Three such mutants have substitutions that rupture the hydrogen bond between aspartate $G1(94)\alpha_1$ and asparagine $G4(102)\beta_2$ that stabilizes the oxy or R-structure (Fig. 7.8), thus biasing the allosteric equilibrium to the T-structure. Replacement of asparagine G10 (108)β by a lysine has the same effect and was the mutation finally selected by Looker et al. (1992).

Before the introduction of gene technology, there had been several successful attempts at linking the αβ-dimers together with bifunctional reagents. For example, reaction of oxyhemoglobin with bis(3,5-dibromosalicyl) fumarate produced a link between lysines $EF6(82)\beta_1$ and β_2 that blocks the diphosphoglycerate binding site (Walder et al., 1980). The oxygen affinity of this complex was as high as that of native hemoglobin stripped of diphosphoglycerate, which made it unsuitable for transfusion. Reaction of deoxyhemoglobin with the same reagent produced a covalent link between lysines $G6(99)\alpha_1$ and α_2 (Chatterjee et al., 1986). This cross-linked hemoglobin reacted with oxygen with a Hill's coefficient of 2.6, but its Bohr effect was reduced to less than half its normal value. The cross-link raised p50 from 6.6 to 13.9mmHg, which is still below the physiologically desirable p50 of 26mmHg.

Reinhold and Ruth Benesch and their coworkers have shown that hemoglobin can be chemically modified in various ways by reaction with derivatives of pyridoxalphosphate. The most promising one has been 2-Nor-2-formyl-pyridoxal 5'-phosphate, which X-ray analysis has shown to link the α-amino group of valine $1\beta_1$ to the ε-amino group of lysine $EF6(82)\beta_2$ (Fig. 7.9). This hemoglobin combines cooperatively with molecular oxygen with a Hill's coefficient of 2.2. Its p50 equals 25mmHg, the same as that of hemoglobin A with diphosphoglycerate, which makes it suitable as a blood substitute (Benesch and Benesch, 1981; Arnone et al., 1977). It now remains to be seen which of the genetically engineered or chemically modified hemoglobins proves least antigenic in medical practice.

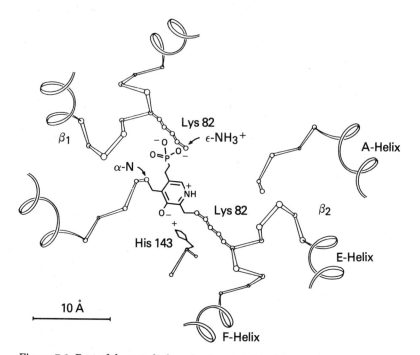

Figure 7.9. Part of the 2,3-diphosphoglycerate binding site, here occupied by 2-Nor-2-formylpyridoxal 5'-phosphate crosslinking valine 1β_1 and lysine EF6(82)β_2. (Reproduced, by permission, from Benesch and Benesch, 1981)

Further Reading

Blow, D. M., A. R. Fersht, and J. Winter, eds. 1986. Design, construction, and properties of novel protein molecules. *Phil. Trans. Roy. Soc. (London) A* **317**:291–457.

Derewenda, U., Z. Derewenda, E. J. Dodson, G. G. Dodson, and Xiao Bing 1991. X-ray analysis of the single chain B29-A1 peptide-linked insulin molecule. *J. Mol. Biol.* **220**:425–433.

Hua, Q. X., S. Shoelson, M. Kochoyan, and M. A. Weiss 1991. Receptor binding redefined by a structural switch in a mutant human insulin. *Nature* **354**:238–241.

Watson, J. D., J. Tooze, and D. T. Kurtz 1983. *Recombinant DNA: A Short Course.* Scientific American. Distr. W. H. Freeman, New York. An introduction to gene technology.

8

From a Tomato Virus to Tumor and Influenza Viruses

Structure of Small RNA Viruses and the Design of Antiviral Drugs

In 1935 Kenneth Smith, an English biologist, discovered a virus in tomato plants and named it *Bushy Stunt Virus* (Smith, 1935). His discovery stimulated two other Englishmen, Fred Bawden at the Rothampsted Experimental Station and Bill Pirie, a young biochemist in Cambridge, to isolate, purify, and characterize the virus. They found that it consisted of only two components: ribonucleic acid and protein. They crystallized it, the first time that a virus had been obtained in truly crystalline form (Bawden and Pirie, 1938). Crystallinity implied that the virus was a molecule, yet it came alive as soon as you rubbed it into the leaf of a tomato plant. This duality blurred the hitherto accepted distinctions between the living and the nonliving world, which induced the radical young Pirie to write an iconoclastic essay entitled "The Meaninglessness of the Terms Life and Living" (Pirie, 1937).

Pirie handed the crystals over to his friend Bernal, who gave them to his American research assistant I. Fankuchen. To measure the size of the virus, Fankuchen had to resolve an X-ray diffracted by a suspension of minute crystals that deviated from the incident beam by only one third of a degree, a difficult feat in those days. With great skill, Fankuchen recorded two powder diffraction rings which Bernal recognized as coming from a body-centered cubic lattice of spheres of 33 nm diameter with a molecular weight of 8–9 million, the largest molecule yet encountered (Bernal et al., 1938).

Forty years later this work culminated in a complete atomic model of the virus protein, illustrating W. L. Bragg's dictum: if you hammer away long enough at great problems, they seem to get tired, lie down, and let you catch them. It lay down for S. C. Harrison at Harvard, after he had hammered away at it for 12 years (Harrison et al., 1978). Eight years earlier J. T. Finch and others in Cambridge, England, had found out that the virus looks like a football (Finch et al., 1970). It has the icosahedral point symmetry 532, which means that it contains 12 fivefold, 20 threefold, and 30 twofold axes of symmetry, causing its surface to be covered by an integral multiple of 60 identical protein subunits arranged in alternate pentagons and hexagons (Fig. 8.1). The coat of tomato bushy stunt virus is made up of 180 identical protein molecules, each consisting of a chain of 387 amino acid residues. The N-terminal 66 residues dip into the interior of the virus. They contain many cationic sidechains apparently evolved to neutralize phosphates of the RNA core. The next 35 residues form a flexible arm. The bulk of the chain is folded into two domains: an inner one that forms the shell of the virus (S) and another projecting outward (P) (Fig. 8.2). Each domain has a characteristic fold of eight β-strands rolled into a barrel (Fig. 8.3). The interior of the protein shell is filled with a randomly coiled chain of RNA containing about 4500 nucleotides. In 1978, solution of this structure was the Everest of protein crystallography, but the impossible of yesterday becomes the commonplace of tomorrow. We now know the structures of other small plant and animal RNA and DNA viruses. Their coat proteins are dissimilar in size and composition, yet X-ray analysis has revealed that nearly all are folded like that of the shell domain in bushy stunt virus, suggesting that these viruses have evolved from one common ancestor (Harrison, 1983; Rossmann and Johnson, 1989). Before X-ray analysis, comparisons of the amino acid sequences of these viruses had provided no hint of their structural similarity. Could it have arisen by convergent evolution, the β-barrel being the only structure that can form patterns of protein subunits distributed symmetrically over a spherical surface? The discoveries of entirely different coat protein structures in a human virus and a small icosahedral bacteriophage (MS2) speak against such a view (Choi et al., 1991; Valegard et al., 1990).

Sindbis Virus belongs to the Toga group of viruses, which are transmitted by arthropods and may cause diseases such as encephalitis, fever, arthritis, and rash. Its core of single-stranded RNA is covered by a spherical shell of 240 protein molecules arranged with icosahedral symmetry. That shell is surrounded in turn by a lipid bilayer spiked by 80 glycoprotein molecules that

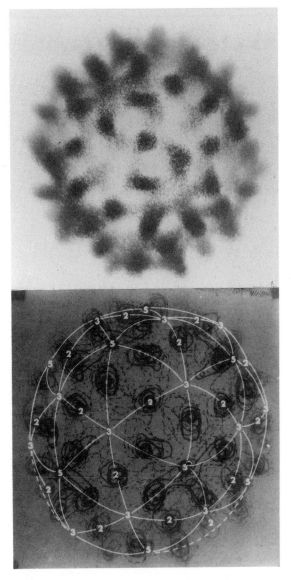

Figure 8.1. (Top) Density plot of an optical reconstruction of negatively stained electron micrographs of tomato Bushy Stunt Virus showing projections of density around five- and sixfold symmetry axes. (Bottom) Contour map of the optical reconstruction with the surface lattice of Caspar and Klug's symmetry group T=3 superimposed (Caspar and Klug, 1962). The numbers mark five-, three-, and twofold axes of symmetry. (Reproduced, by permission, from Crowther, 1971)

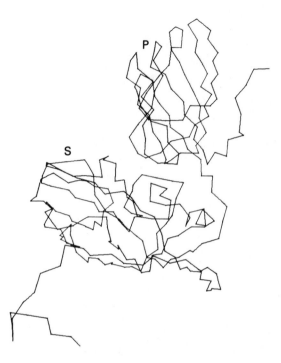

Figure 8.2. Trace of α-carbon positions in the coat protein of tomato bushy stunt virus, the first virus structure to be solved. The coat is made up of 180 such molecules. P stands for projecting and S for shell domain. (Reproduced, by permission, from Harrison et al., 1978)

recognize host cell receptors. The viral genome encodes two long polypeptides that are cleaved into smaller proteins by viral and cellular proteinases. One of these proteinases is the core protein which cleaves itself autocatalytically from its precursor, after which its catalytic activity stops.

Figure 8.3. (opposite) (A) Schematic drawing of the polypeptide fold found in the coat protein of most small spherical RNA viruses. Tomato bushy stunt virus lacks the two helices, represented as cylinders, that are present in polio virus. (Drawing by Jane Richardson reproduced, by permission, from Hogle et al., 1985) (B) Polypeptide fold of Sindbis Virus coat protein. The arrows mark β-strands which are labeled A, B, C, etc., from the amino ends of the two domains. H141, D163, and S215 mark the catalytic triad histidine, aspartate, serine. Note the C-terminal tryptophan (C264) blocking the catalytic site. (Reproduced, by permission, from Choi et al., 1991)

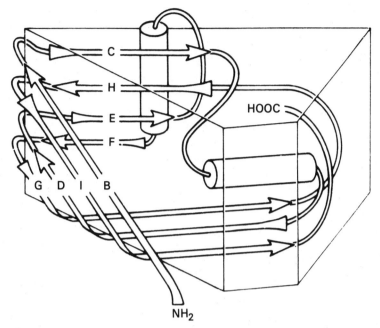

A

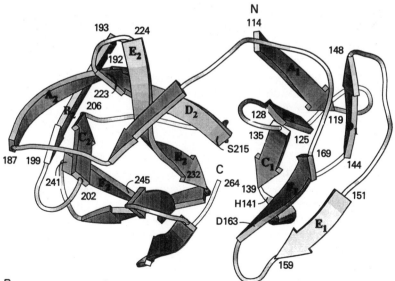

B

Michael Rossman's group at Purdue University crystallized the core protein as a naked and hollow shell and determined its structure by X-ray analysis. It is made up of two identical polypeptides of 264 amino acid residues each; their first 113 residues are rich in prolines and basic residues; they are disordered and therefore invisible in the crystal. In the virus they probably reach into the interior of the shell to bind the RNA. The remaining protein resembles the mammalian serine proteinases of the chymotrypsin family which consist of two β-barrels flanking a catalytic cleft (See Chapter 2, page 64) In that cleft, peptides are cleaved by a triad of an aspartate, a histidine, and a serine. The carbonyl group of the peptide substrate is hydrogen-bonded to, and polarized by, the amino group of a glycine; sidechain specificity is provided by a pocket that flanks the catalytic site. With a single exception, all these residues in and around the active site found to be invariant in mammalian serine proteinases are present also in this viral core protein, and their atomic coordinates differ by an average of only 0.8Å from those of the chymotrypsin-like bacterial α-lytic protein-ase. In the core protein the active site is blocked by the protein's own C-terminal tryptophan whose indol sidechain occupies the specificity pocket and whose carbohydrate faces the active serine and histidine (see Fig. 2.16 on page 65). Having cleaved itself from its polypeptide precursor, the enzyme has killed itself to become a passive coat protein. Other viral genomes also code for serine proteinases, but Sindis virus is the first one found to convert that proteinase into the genome's envelope; it is a remarkable example of Nature's economy and of the conservation of protein structures in divergent forms of life (Choi et al., 1991).

The *Picornaviruses* (small RNA viruses) include entero-, rhino-, cardio-, and aphthoviruses. The structure of at least one member of each of these groups has been solved: the poliovirus, the common cold virus, the murine encephalomyocarditis (Mengo) virus, and the foot and mouth disease virus. The coats of all these viruses are made up of 60 copies of each of three different viral proteins (VP 1–3). Their secondary structures conform to the same pattern that was first found in tomato bushy stunt virus: β-strands wrapped around a β-barrel. While the coat of the tomato bushy stunt virus contains 180 identical protein molecules, with one pair of molecules flanking each of the 30 true twofold axes and each of the 60 pseudo-twofold axes of symmetry, the 60 copies of each of the three proteins of the picornaviruses are arranged so that five, three, or two copies of each protein are grouped around the five-, three-, and twofold axes of

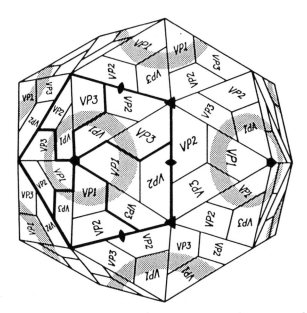

Figure 8.4. Distribution of the three coat proteins, (VP1, 2, and 3) on the 20 faces of the icosahedral picornaviruses. The shaded patches mark the canyons that probably serve as receptor sites. ◆ *— fivefold,* ▲ *— threefold, and* ◗ *— twofold symmetry axes. (Reproduced, by permission, from Giranda et al., 1990)*

symmetry (Fig. 8.4). The three proteins differ in the structures of their terminal extensions and of the loops connecting their β-strands. Their N-terminal extensions serve as arms that embrace the neighboring subunits and knit the proteins together into stable spherical shells. Sixty copies of a fourth, smaller protein (VP4) of different structure lie underneath that shell (Fig. 8.5) (Kim et al., 1989; Arnold and Rossmann, 1990; Rossmann and Johnson, 1989).

The serological specificity of the virus proteins is determined by the patterns of amino acid residues on their external loops, which tends to change in response to evolutionary pressure. Most antigenic mutants of the poliovirus that resisted neutralizing monoclonal antibodies exhibited amino acid substitutions on exposed loops of the viral proteins. Substitutions at less exposed sites that caused resistance often did so because they destabilized the conformation of an exposed loop. For example, aspartate 2164 stabilizes an exposed loop because its carboxylate forms hydrogen bonds with an NH groups in the opposite strand of the loop. Replacement of the

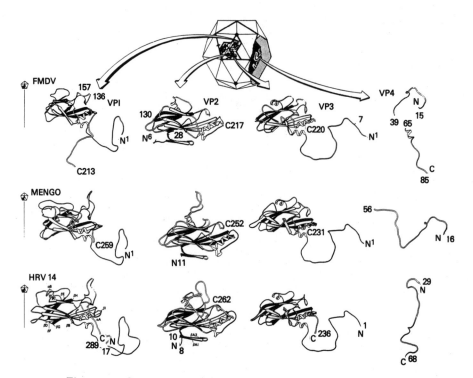

Figure 8.5. Comparison of the structures of the coat proteins VP1, VP2, and VP3 and the internal protein VP4 of the foot and mouth disease virus (FMDV), the meningoencephalitis virus (MENGO), and the human rhinovirus (HRV). Note the similar structures of the β-barrels and the marked differences between the structures of some of the external loops that determine the antigenicities of the viruses. (Reproduced, by permission, from Acharya et al., 1989)

aspartate by either asparagine or histidine breaks these bonds and thus makes the virus resistant to neutralizing monoclonal antibodies directed against that loop (Fig. 8.6) (Page et al., 1988). The receptor-binding sites of the virus proteins often reside in pits or canyons lined with invariant residues that recognize specific receptor molecules and are too narrow to allow antibodies access (Figs. 8.4, 8.7A and B, pages 195 and 196). The structure of that canyon in the rhinovirus suggests that infection might be prevented by compounds that block it. This avenue has not yet been explored, but Rossmann and his colleagues who solved the structure of the rhinovirus have discovered another promising approach. Chemists at the Sterling-Winthrop Research Institute had synthesized a series of structurally

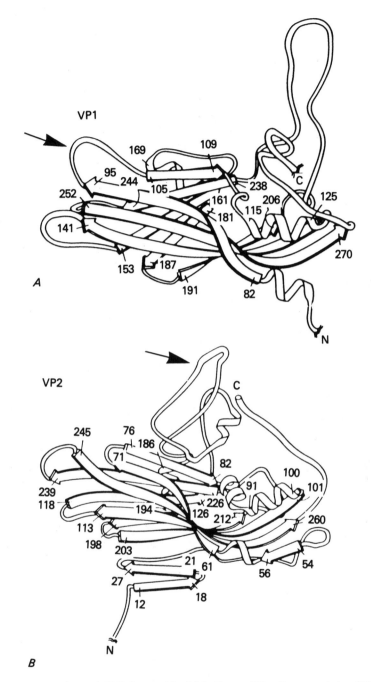

Figure 8.6. (A and B) Polypeptide fold of two of the three proteins (VP1 and VP2) forming the surface coat of the poliovirus. The folds are similar to those of most other small spherical RNA viruses. The arrows mark the exposed loops referred to in the text.

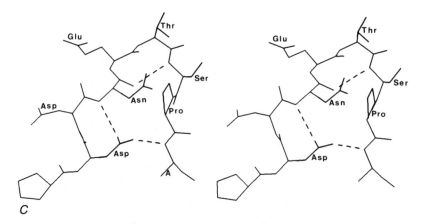

C

Figure 8.6. (C) Detailed structure of the exposed loop marked by the arrow in (B), showing residues from proline 2163 on the left to proline 2170 on the right. Replacement of aspartate 2164 by asparagine or histidine allows the virus to escape neutralizing monoclonal antibodies, apparently because the hydrogen bonds made by the aspartate (broken lines) are needed to maintain the specific conformation of the loop that antibodies recognize. (Reproduced, by permission, from Page et al., 1988)

related antiviral compounds that inhibit replication of the rhinovirus and other picornaviruses by preventing the shedding of their protein coats. Rossmann and his colleagues have cocrystallized several of these compounds with the rhinovirus and determined their binding sites by X-ray analysis. The rhinovirus protein contains a largely hydrophobic cavity connected to the surface by a narrow pore. All the compounds tested were found in that cavity, elbowing some of the surrounding amino acid residues aside and displacing two water molecules (Fig. 8.8). Their interactions with the protein were predominantly hydrophobic. As in hemoglobin, the exact mode of binding depended critically on the stereochemistry of the drug and the binding site. The orientations of two drugs are reversed, yet they differ merely by the lengths of their aliphatic chains and by the additional methyl group on WIN 52084 (Fig. 8.9, page 198). The inhibitory action of WIN 52084 also varies with the orientation of the methyl group relative to the oxazoline ring. The S isomer is 10 times more active than the R isomer; in fact, the latter binds so weakly that it could not be located by X-ray analysis. These are further examples of the rule found in drug binding to hemoglobin, discussed in Chapter 4, that minor alterations in the stereochemistry of drugs can induce major changes in the way they bind to proteins.

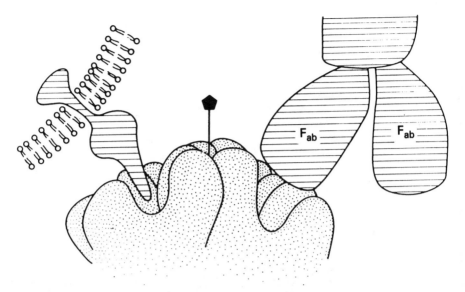

The Canyon Hypothesis

Figure 8.7. (A) Surface of five viral coat proteins grouped around a symmetry axis, marked by the black pentagon. The cellular receptor protein on the left docks in a narrow canyon that the antibody on the right cannot penetrate. The base of the canyon is lined with largely invariant amino acid residues recognized by the receptor, while its rim is surrounded by polypeptide loops of variable sequence that can escape recognition by antibodies. (Reproduced, by permission, from Rossmann and Palmenberg, 1988)

Rossmann's group also investigated the structure of two drug-resistant mutants of the rhinovirus and found their resistance to be due to the replacements of amino acid residues in the drug-binding cavity. In one mutant valine 1188 was replaced by leucine and in the other cysteine 1199 by tyrosine. In each case the more bulky sidechain offered steric hindrance to the binding of the drugs. Mutants in which valine 1188 is replaced by methionine, or cysteine 1199 by tryptophan or arginine, were also resistant for the same reason (Smith et al., 1986; Rossmann, 1989; Badger et al., 1989).

The epithelial receptor for most rhinovirus serotypes appears to be a cell surface molecule named *Intercellular Adhesion Molecule 1* or ICAM1 (Staunton et al., 1989). Its amino acid sequence suggests that it contains five immunoglobulin-like domains, like the neural

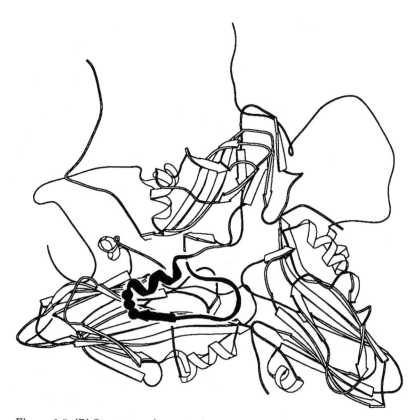

Figure 8.7. (B) Structure of part of the surface of foot and mouth disease virus, showing the proteins VP1, VP2, and VP3 grouped around a pseudo-threefold axis of symmetry. The black circles mark the positions of Arg145-Gly145-Asp147 on a surface loop of VP1. This tripeptide appears to be essential for recognition of the receptor. Either alone or as part of the surface of the lens protein α-crystallin, this tripeptide inhibits binding of the virus to its receptor. (Courtesy of Dr. David Stuart)

cell adhesion molecules described on p. 59. Mutational analysis identified the N-terminal domain as the receptor or at any rate part of the receptor moiety. A molecular model of this domain can be made to fit into a model of the rhinovirus' canyon such that the charges of its buried ionized sidechains are compensated and residues whose replacement inhibits binding are part of the contact area. By itself, the N-terminal receptor domain does not fill the canyon, which suggests that additional contacts might be made with a second domain (Giranda et al., 1990).

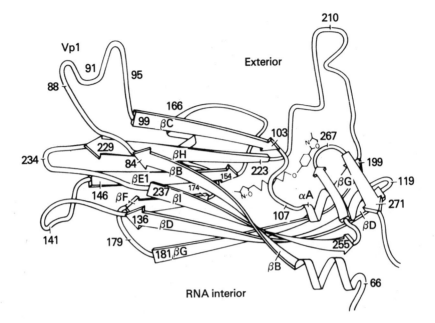

Figure 8.8. Ribbon drawing of the polypeptide fold of one of the three coat proteins of the human rhinovirus, showing an antiviral compound docked in its cavity. (Reproduced, by permission, from Smith et al., 1986)

Engineering a Better Polio Vaccine

Sabin's live attenuated vaccine has almost eliminated polio in developed countries. It contains all three poliovirus serotypes. Type 1 is safe because it contains multiple attenuating mutations, but type 3 differs from the wild type by only two mutations, which implies a finite probability of reversion to wild type. In consequence type 3 can give rise to infections even though this happens very rarely. The same also applies to type 2.

J. W. Almond and his colleagues at Reading, England, have used J. M. Hogle's atomic model of type 1 to engineer a chimeric virus that combines the antigenicity of type 3 with the multiple mutations of type 1. One of the major antigenic sites of type 3 is an external loop containing residues 89–100 of viral protein 1 (VP1) (See arrow in Fig. 8.6A on page 193). Almond's team replaced eight of these amino acid residues in VP1 of the Sabin strain by the corresponding residues in type 3 (-Ser-Ala-Ser-Thr-Lys-Asn-Lys-

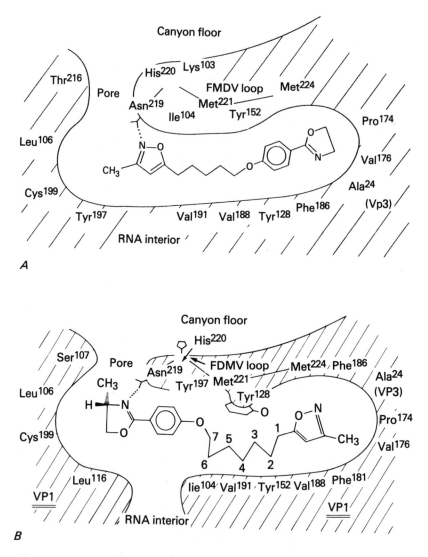

Figure 8.9. (A) Docking of the antiviral compound WIN VI with a hydrogen bond to aspartate 219. (Reproduced, by permission, from Badger et al., 1989) (B) Docking of the antiviral compound WIN 52084 in reverse orientation. Note the hydrophobic lining of the canyon. (Reproduced, by permission, from Smith et al., 1986)

Asp- by -Glu-Gln-Phe-Thr-Thr-Arg-Val-Gln-) (Burke et al., 1988; Minor et al., 1990).

The chimeric virus was assayed by inoculating it with pairs of antisera, each of which is specific for two types of polio virus, i.e., for 1 and 2, 2 and 3, or 1 and 3. All three pairs neutralized the chimeric virus; the virus was also neutralized by type 1 or type 3 antiserum alone, but not by type 2 antiserum. The results show that the chimera has antigenic characteristics of both type 1 and type 3, but is not a mixture of these two types. A similar experiment has been done by Murray and others (1988). Martin and others replaced the six residues in the same loop of VP1 by the corresponding residues in the Lansing strain of VP 2 that produces paralytic disease in mice. When injected into rabbits, this chimera of VP1 and 2 induced a neutralizing response to both strains of virus; moreover it proved to be highly neurovirulent in mice (Martin et al., 1988). It is remarkable and frightening that so small a change should be sufficient to induce susceptibility in an animal species immune to type 1.

Structure of Tumor Viruses

The β-barrel structure shown in Figure 8.3 on page 189 that is common to the coat proteins of many small RNA viruses has also been found in three DNA viruses: in polyoma virus which causes tumors in mice, in simian virus 40 (SV40) which causes tumors in hamsters, and in adenovirus which causes mainly respiratory diseases in man, but also causes tumors in some rodents. Polyoma virus and SV40 have similar structures. Electron micrographs showed their shells to be icosahedra of about 500Å diameter, covered by 72 protein units.

In a spherical shell with icosahedral symmetry, covered by 72 protein units, 12 units must lie on the axes of fivefold symmetry. The remaining 60 units must be placed between the symmetry axes, at positions where each unit is surrounded by six others at the corners of a regular hexagon (Fig. 8.10). In a protein shell made up of identical subunits, contacts between subunits should be equivalent or nearly so (Caspar and Klug, 1962). This rule requires a protein unit that is surrounded by six others to be made up of six subunits. Accordingly, the protein shells of *polyoma virus* and *SV40* were expected to be made up of pentagons at the 12 fivefold positions plus hexagons at the remaining 60 positions, making 420 subunits in all. On the other hand, chemistry indicated a smaller number of

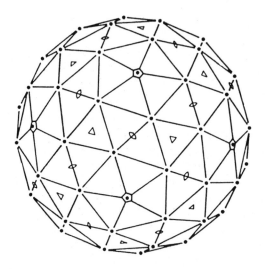

Figure 8.10. Arrangement of symmetry elements and coat protein pentamers in the icosahedral shells of simian virus 40, murine polyoma virus, and related viruses. The symbols ◇, △ , and ○ mark five-, three-, and twofold symmetry axes. The circular dots mark the positions of the 12 pentamers on the fivefold axes and of the 60 pentamers between the symmetry axes.

subunits. The contradiction was resolved when X-ray analysis showed that nature has broken Caspar and Klug's equivalence rule.

The structural units turned out to be all pentagons, making a total of 5 × 72= 360 protein subunits, in agreement with the chemical evidence; instead of being equivalent, the contacts between pentagons are of three different kinds, depending on their positions (Fig. 8.11) (Rayment et al., 1982). Each subunit is folded into a β-barrel whose N-terminal arm dips into the interior of the viral shell, while its C-terminal arm embraces a neighboring pentamer and helps to knit the protein shell together (Fig. 8.12). Each virus particle also contains 72 copies of each of two other proteins that are probably internal. Its core contains a minichromosome of a 5.3 kilobase DNA organized into 25 nucleosomes by four different histones (Liddington et al., 1991; Yan et al, 1990; Harrison, 1990).

Adenovirus is an icosahedron of about 900Å diameter. Its faces are covered by 240 protein units known as hexons, while its 12 corners are occupied by two other proteins; one forms a pentameric base and the other is a long spike with a knob at its tip (Fig. 8.13) (Valentine and Peirera, 1965). Each hexon contains three polypeptide

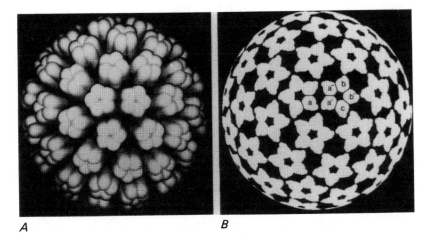

A   B

Figure 8.11. (A) Three-dimensional reconstruction of the viral surface. (B) Diagram of the different contacts at the points a and a′ where three pentamers meet, and at b and b′, and c and c′ where two pentamers meet. (Reproduced, by permission, from Salunke et al., 1986)

chains of 967 amino acids. The hexons have been isolated and crystallized, and their structure has been determined by X-ray analysis. Each chain is folded into two β-barrels. Two of the loops connecting the barrels are very long and are folded into a tower that protrudes from the viral surface (Fig. 8.14). These two towers and the heads of the spikes contain some of the viral antigens (Van Oostrum and Burnett, 1985).

The structures of the coat proteins of the three tumor viruses bear no direct relationship to their oncogenic action, because the proteins that induce malignancy do not form part of the viral coats. Malignancy of the polyoma virus is due to three proteins known as tumor or T-antigens that are produced early in the virus's replicative cycle, before the coat proteins are synthesized. When the gene for one of these T-antigens was spliced into the chromosome of the Rous sarcoma virus in place of its own oncogene, the T-antigen transformed chicken fibroblasts to malignancy. This was found to be due to its binding to a chicken tyrosine kinase that forms part of the fibroblast's control of cell division. Combination with T-antigen inhibits the phosphorylation of one of the kinase's own tyrosines, and this inhibition in turn puts the kinase's normal constraint on cell division out of action (Markland and Smith, 1987).

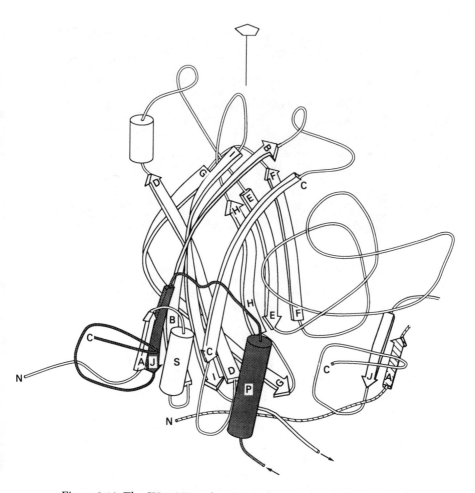

Figure 8.12. The SV40 VP1 subunit. The strands of the β-barrel are indicated
by letters A–J. The pattern of connections in strands B–I is the same as in
the RNA virus subunits (Fig. 8.3, page 189). Strand J comes from the
invading C-terminal arm of another pentamer. The helix marked P is
important for contacts between pentamers. It interacts with two other P
helices in the threefold pentamer clusters (contact a in Figure 8.11) and
with one other P helix in one of the types of twofold cluster (contact b in
Figure 8.11). In the other type of twofold cluster (contact c in Figure 8.11), it
becomes an extended strand, interacting with the A-strand of the β-barrel.
The N-terminal arm (A—N) from the neighboring subunit is shown hatched
on the bottom right. (Reproduced, by permission, from Harrison, 1990)

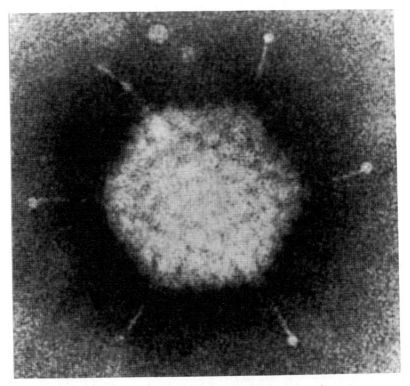

Figure 8.13. Negatively stained electron micrograph of the adenovirus, showing the hexons on the faces and the pentons with their spikes at the corners of the icosahedron. (Reproduced, by permission, from Valentine and Pereira, 1965)

The oncogenic action of adenovirus and simian virus 40 came to be understood by a study of a rare inherited malignancy known as retinoblastoma. This is a recessive disease due to deletion or mutation of another gene involved in the control of cell division. The product of this gene, which is needed to *prevent* retinoblastoma, is called, misleadingly, *retinoblastoma protein* or *RB*. Adenovirus and simian virus 40 each produce a different protein that associates with RB and puts it out of action. As a result, control of cell division is impaired and malignancy follows (White et al., 1988; De Caprio et al., 1988). The genes for the fibroblast's normal cellular kinase and for RB have become known as anti-oncogenes. The three viruses induce malignancy by synthesizing antagonists to their products; they contain anti-anti-oncogenes.

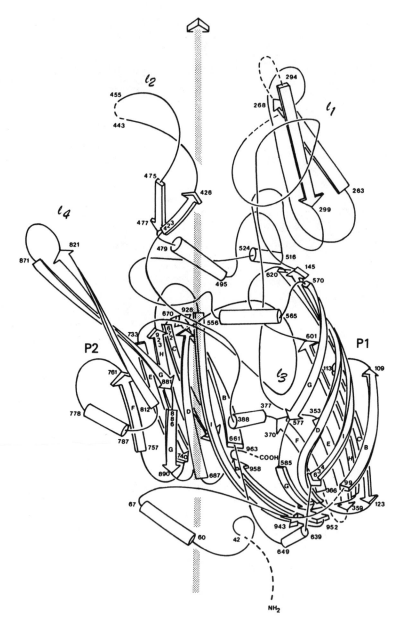

Knowledge of the structure of the coat proteins of tumor viruses may not help to control their oncogenic action directly, but it may help the development of antibodies or drugs that inhibit the first step, the viruses' attachment to a cellular receptor or their uncoating. As of 1991, the medically more important structures of the protein products of the oncogenes, anti-oncogenes, and anti-anti-oncogenes are still unknown.

The GTP-ase of the *Ras* Oncogene

In 1964 J. J. Harvey in London injected into new-born mice cell-free plasma from a rat infected with Moloney's leukemogenic virus. When several of the mice developed tumors at or near the injection site, instead of leukemia, Harvey attributed these to a new, hitherto unobserved virus (Harvey, 1964). Three years later W. H. Kirsten and L. A. Mayer in Chicago found that cell-free plasma of rats infected with an erythroblastosis virus induced sarcomas in mice; they noted that the causative virus had properties resembling those described by Harvey (Kirsten and Mayer, 1967). These two related, but distinct, pathogens have become known as the Harvey and the Kirsten murine sarcoma viruses. Their genomes are made up of portions of RNA of the virus from which they were originally derived, combined with portions of endogenous rat RNA. On its own, neither virus can replicate, but DNA transcripts of their genomes are capable of being spliced into the chromosomes of normal mouse fibroblasts in culture and of transforming them into malignant ones. They can also induce sarcomas and erythroblastomas in susceptible mice.

In 1979 Skólnik and his coworkers found that fibroblasts transformed by Harvey-Kirsten viruses contained substantial concentrations of an apparently new protein of 21,000 daltons encoded by the viral RNA. They called it p21. Suspecting that it might be the protein responsible for the transformation, they searched for and isolated a mutant of the virus that was temperature-sensitive

Figure 8.14. (opposite) Fold of the 963 amino acid residues–long polypeptide chain in the hexon coat protein of the adenovirus. The bottom of the molecule faces the interior, and the tower T, made up of the loops l_1, l_2, and l_4, faces outward. The N-terminus is at the bottom and the C-terminus slightly above it. The globular part of the chain is folded into two β-barrels, marked P1 and P2. Three such chains constitute one hexon. (Reproduced, by permission, from Roberts et al, 1986)

for the maintenance of several properties of the transformed fibroblasts. The p21 protein encoded by this mutant proved indeed to be more temperature-labile than p21 encoded by the wild-type virus. This was the first demonstration that a single protein can transform normal cells into proliferating ones. (Shih et al., 1979a).

It then emerged that a wide range of normal mammalian cells contain a protein closely related to the viral p21 protein, but present only at low concentration (Ellis et al., 1981). When this same normal protein was synthesized in fibroblasts in sufficiently high concentration, it proved capable of transforming them. The crucial advance came in 1982, when several groups of workers discovered independently that human bladder carcinomas contained a mutated allele of the normal cellular gene for the p21 protein (Tabin et al., 1982; Reddy et al., 1982; Taparowsky et al., 1982). The mutation consisted of a single base change, resulting in the substitution of a valine for a glycine in position 12 of the p21 polypeptide chain. This glycine has since been found to be essential for the normal cellular function of the p21 protein.

How could a single amino acid substitution render a normal protein carcinogenic? A combination of biochemical and crystallo-graphic work has gone some way toward providing the answer, but first the notation now generally adopted for this class of proteins needs to be introduced. The genes coding for them are known as *ras* (for rat sarcoma). The three *ras* genes so far found in the mammalian genome are called H- or Ha-*ras* (for Harvey), K- or Ki-*ras* (for Kirsten), and N-*ras* (for symmetry reasons: *Hij*K*lm*N). The proteins are called $p21^{ras}$, and specifically, $p21^{Ha-ras}$, $p21^{Ki-ras}$, and $p21^{N-ras}$. Sometimes the prefix "c" is added for the normal cellular gene or protein and "v" for the viral ones. The viral genes are known as oncogenes and their normal cellular counterparts as proto-oncogenes.

All cellular *ras*-proteins catalyze the hydrolysis of GTP to GDP and inorganic phosphate. They are partly homologous to a large class of GTP-ases now known as the G-proteins. (For reviews see Weinberg, 1985; Barbacid, 1987; Bourne et al., 1991). They are membrane-bound enzymes that work as molecular switches for the transmission of a variety of signals to other proteins that bind to them. They can exist in three states: the oncogenically active one combined with GTP, the inactive one combined with GDP, and a transient, inactive empty state (Fig. 8.15).

The crystal structures of both the active and the inactive states of p21 have been solved. Pai and others have determined the

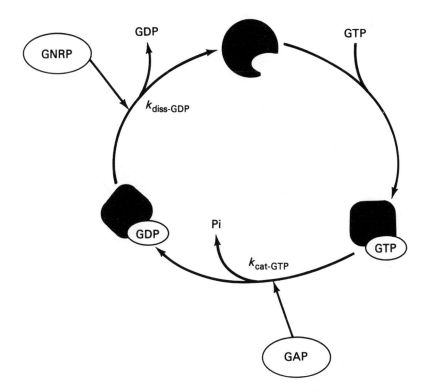

Figure 8.15. The GTPase cycle, which is regulated by GAP (meaning GTPase-activating protein), and GNRP (meaning guanine-nucleotide release protein). $k_{cat\text{-}GTP}$: rate of GTP hydrolysis; $k_{diss\text{-}GDP}$: rate of GDP dissociation from GTPase. (Reproduced, by permission, from Bourne et al., 1991)

structure of the nucleotide-binding domain of the normal cellular p21 H-*ras* protein with GTP (Pai et al., 1989) (Fig. 8.16). It turned out to be similar to that of another GTP-binding protein determined earlier: the bacterial elongation factor Tu, which provides part of the energy for the addition of amino acid residues to the growing polypeptide chain in the ribosome (La Cour et al., 1985). The structures are made of a core of twisted β-strands (β_1–β_6) forming a pleated sheet. This is flanked by α-helices (α_1–α_5). The β-strands and α-helices are linked by external loops (λ_1–λ_{10}). Mg-GTP binds in a superficial groove that extends over the entire width of the protein from loops λ_4 to λ_8 and λ_{10}. The nucleotide is held there mainly by a complementary constellation of hydrogen bond donors

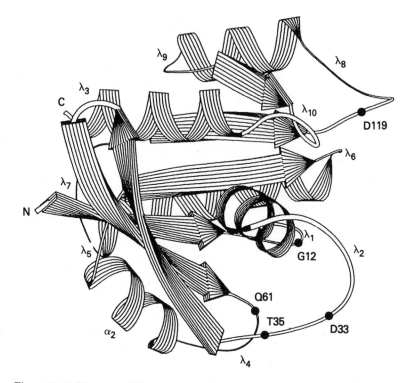

Figure 8.16. Diagram of the structure of the nucleotide-binding domain of the normal human cellular protein c-Ha-ras p21. G- glycine; D- aspartate, T- threonine, Q- glutamine. (Reproduced, by permission, from Pai et al., 1989)

and acceptors, while the magnesium ion is coordinated to six oxygens at the corners of an octahedron: one each from the β- and γ-phosphates of GTP, one from a serine, another from a threonine OH, and two oxygens from bound water molecules (Fig. 8.17).

Milburn and others have solved the structure of the oncogenically inactive, GDP-bound form of the enzyme (Milburn et al., 1990). Comparison of the two structures shows that hydrolysis of GTP is accompanied by a contraction of the active site, brought about by inward movements of loops λ_2 and λ_4 and a shift of helix α_2. When all the α-carbon atoms except the ones in those three segments are superimposed, substantial shifts of the α-carbons in those three segments relative to those in the rest of the molecule were found: 2.6 ± 0.1Å in loop λ_2, 3.6–7.4Å in loop λ_4, and 1.3–2.3Å in helix α_2.

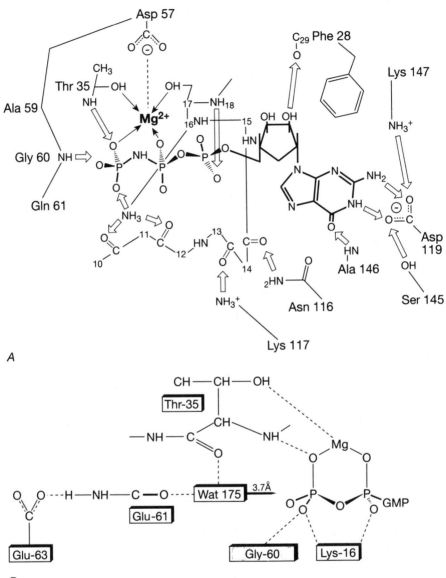

Figure 8.17. (A) Interaction between the GTP analogue and the surrounding c-Ha-ras p21 protein. Open arrows ⇑ *indicate hydrogen bonds. Solid arrows stand for electrostatic interactions with Mg^{2+}. (B) Coordination of the β- and γ-phosphate of GTP, deduced from the structure of its analogue, and of the water molecule 175, indicating the direction of that molecule's possible nucleophilic attack on the phosphorus atom. (Reproduced, by permission, from Pai et al., 1989)*

Shifts in the same direction, but smaller, were also found by new X-ray techniques that allowed the reaction of the enzyme with GTP to be followed in a single crystal. These shifts alter the affinity of the *ras*-protein for its receptors, thereby enabling it to act as a molecular switch. In the past, allosteric proteins whose activities change in response to chemical stimuli have been found to be made up of several polypeptides arranged in two alternative quaternary structures. G-proteins, on the other hand, are monomers that alter their structure by induced fit, rather like the heat-shock protein described in on page 61 or yeast hexokinase which closes its jaws on addition of glucose to bring its two substrates, glucose and ATP, into juxtaposition (Stryer, 1988).

Point mutations causing the replacement of single amino acid residues sufficient to turn *ras* proto-oncogenes into oncogenes have been found in the coding triplets for glycines 12 and 13, alanine 59, and glutamine 61 (Barbacid, 1987). All these replacements impair the GTP-ase activity and shift the equilibrium between the two states of the enzyme toward the oncogenically active, GTP-bound form. Glycines 12 and 13 are essential for catalysis because they are part of loop λ_1 which makes such close contact with the phosphates and the effector-binding loops λ_2 and λ_4 that sidechains cannot be fitted in without disrupting the structure. Glutamine 61 is hydrogen-bonded to a water molecule (175) that is believed to act as a nucleophile in the hydrolysis of the β–γ phosphate bond; the glutamine sidechain is needed to hold that water molecule in position. Replacement of glycine 12 by valine, as in the oncogene of the human bladder carcinoma, pushes glutamine 61 and water molecule 175 apart, which weakens the water's nucleophilicity and therefore slows down catalysis. Replacement of glycine 12 by an arginine, also found in oncogenic *ras* proteins, displaces water 175. Alanine 59 lies close to the γ-phosphate of GTP; its replacement by a larger residue may shift either the phosphate or the loop λ_4 of which the alanine forms part (Bourne et al., 1991) (Fig. 8.18). Mutations of the *ras* gene have been found in two-thirds of human colo-rectal cancers, most of them at position 12 of the cellular Kirsten gene (Bos et al., 1987).

Purified normal p21^{ras} hydrolyzes GTP very slowly. *In vivo* catalysis is accelerated by a cofactor, GTPase activating protein (GAP), that makes contact with the "effector loops" λ_2 and λ_4. X-ray analysis shows that the loop λ_4 that carries the essential glutamine 61 is rather loose. It has been suggested that combination with the activating protein may immobilize the loop, thereby fixing water 175

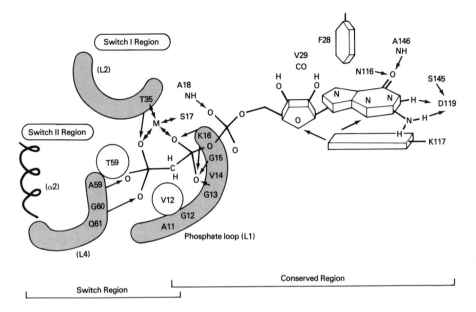

Figure 8.18. Schematic drawing of the interactions between GTP and protein, showing the critical position of loops λ_1, λ_2, and λ_5; also the effects of substituting valine for glycine in position 12 and threonine for alanine in position 59. (Reproduced, by permission, from Milburn et al., 1990)

more firmly in position and accelerating catalysis. Conversely any mutation that weakens interaction between the two proteins would shift the equilibrium of the *ras* protein toward a GTP-bound, oncogenically active, but catalytically inactive form. That form is believed to activate, either directly or indirectly, a cellular growth factor and thus induce malignancy.

The complete chain of molecular events leading to the activation of growth factors has not yet been unraveled. Nevertheless, the discovery of the *ras*-genes, the isolation of the *ras*-proteins, the discovery of the single amino acid substitutions that render them oncogenic, and finally the unravelling of the stereochemical mechanism of that transformation have provided deeper insights into carcinogenesis than any previous set of experiments.

Influenza Virus

The structure of the influenza virus is different from those of the viruses discussed earlier in this chapter. Its core is made up of several strands of RNA surrounded by a phospholipid membrane. Anchored in this membrane are two stalk-like proteins: the hemaglutin whose structure was determined by Don Wiley and his colleagues at Harvard together with John Skehel and his colleagues in London (Wiley and Skehel, 1987), and the neuraminidase whose structure was determined by P. M. Colman and others in Canberra, Australia (Varghese et al., 1983; Colman et al., 1983).

The hemaglutin molecule is a trimer. Each of its monomers is made up of two chains, one containing 328 and the other 221 amino acid residues, linked together by a disulphide bridge. The C-terminal portion of the longer chain contains a hydrophobic membrane anchor that was enzymatically removed to make the hemaglutin soluble for crystallization. This soluble hemaglutin molecule contains three distinct segments: a region of pleated sheet near the membrane anchor, a long, partly helical stalk that holds the three monomers together, and a globular head that contains an eight-stranded β-barrel of topology similar to that of the coat proteins described earlier, with antigenic loops protruding from its surface. The β-barrel carries the receptor-binding site that attaches the virus to a carbohydrate, sialic acid, on the surface of its host's receptor cell (Figs. 8.19 and 8.20).

How can the virus defy the neutralizing antibodies built up in previous infections and yet always bind to the same sialic acid receptor? Wiley and Skehel found the sites of antigenic variations to be distributed over five separate areas on the hemaglutinin surface. One of these areas surrounds the receptor binding site; antibodies covering that area would shield it from the receptor. Conversely, if the binding of antibodies were prevented by anti-genic variation, the receptor-binding site would be laid bare. That site forms a harbor or canyon narrowly walled in to make it inaccessible to antibodies, yet accessible to the receptor. As in the rhinovirus, it is lined with an invariant constellation of amino acids; they tie the sialate to its moorings by hydrogen bonds (Fig. 8.21). Knowledge of the stereochemistry of the binding site offers the chance of designing drugs that block it and prevent the receptor from docking there (Skehel et al., 1989).

The neuraminidase is an enzyme that cleaves the terminal sialic acid from glycoconjugates on cell surfaces, so that the virus can be released from infected cells. It is a tetramer made up of four globular subunits at the end of a long stalk anchored in the viral membrane (Fig. 8.22). Each of the subunits is made up of six structurally identical β-sheets arranged like the blades of a propeller (Fig. 8.23). Its catalytic site that combines with and cleaves off sialic acid lies in

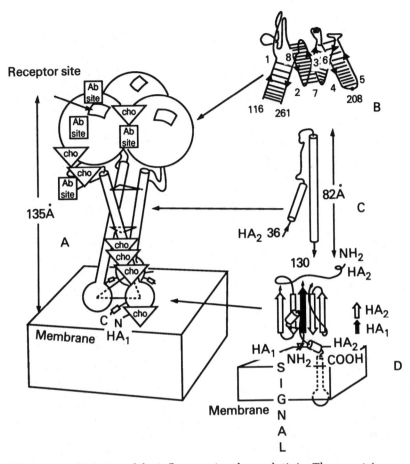

Figure 8.19. Structure of the influenza virus hemaglutinin. Three protein molecules are grouped around a symmetry axis. They are held together by their long α-helices, indicated by the long cylinders, which form a triple-stranded coiled coil. At the top is a globular domain made up of an eight-stranded pleated sheet connected by irregular loops. Ab-site: antibody combining sites; cho: glycosylation sites. The proteins are anchored in the viral membrane by a hydrophobic peptide indicated by the dotted bottle-shape at the bottom right. (Reproduced, by permission, from Wiley and Skehel, 1987)

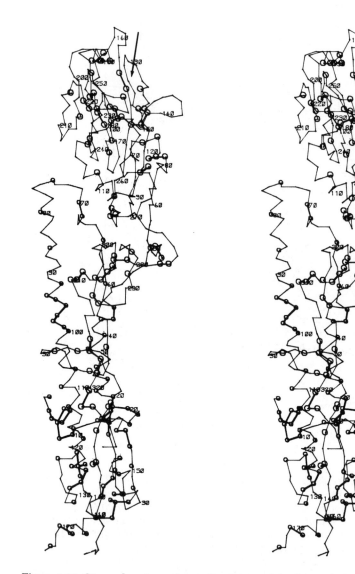

Figure 8.20. Stereo drawings of α-carbon trace of the hemaglutinin. The circles mark residues that remain the same in different virus strains. (Courtesy of Dr. D. C. Wiley. See also Wilson et al., 1981.)

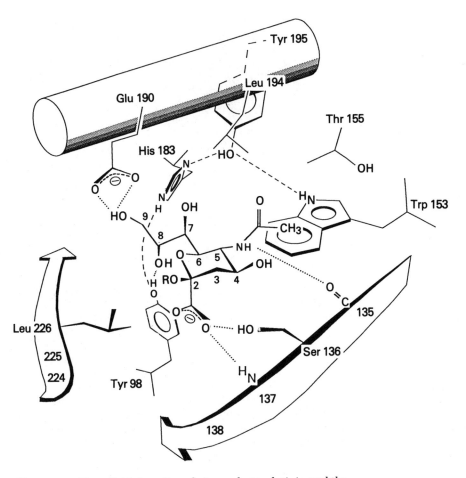

*Figure 8.21. Potential interactions between hemaglutinin and the
α-N-acetylneuraminic acid moiety of the receptor analogue sialyllactose
used in the crystallographic analysis. Note the many possible hydrogen
bonds. (Reproduced, by permission, from Weis et al., 1988, and Sauter
et al., 1989)*

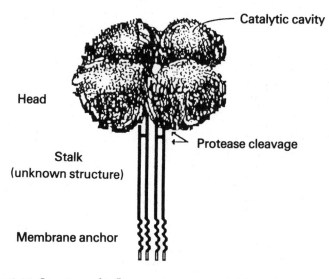

Figure 8.22. Structure of influenza virus neuraminidase: the assembled tetramer. Four globular heads are perched on a stalk that carries a hydrophobic membrane anchor. (Reproduced, by permission, from Air et al., 1989)

a deep crater on its top surface and is lined with invariant amino acids. As in the hemagglutinin, the burying of the catalytic acid site in a deep and narrow cleft makes it inaccessible to antibodies. "Escape mutants" that fail to react with antibodies against the influenza virus were found to have single amino acid replacements on the outer surface of the neuraminidase. The structure of the neuraminidase opens further avenues toward the development of antiviral drugs (Air et al., 1989; Tulip et al., 1991; Varghese and Colman, 1991).

Further Reading

Barbacid, M. 1987. *Ras* genes. *Ann. Rev. Biochem.* **56**:779–828.

Caspar, D. L. D. and A. Klug 1962. Physical principles in the construction of regular viruses. *Cold Spring Harbor Symposium Quant. Biol.* **27**:1–26.

Mittnacht, S., D. Templeton, and R. A. Weinberg 1991. Functioning of the retinoblastoma gene. Accomplishments in Cancer Research 1990, pp. 197–203. J. B. Lippincott Co., Philadelphia, PA.

Rossmann, M. G. and J. E. Johnson 1989. Icosahedral RNA virus structure. *Ann. Rev. Biochem.* **58**:533–574.

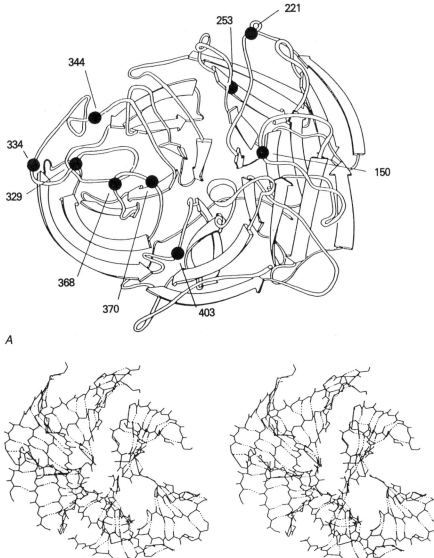

Figure 8.23. (A) Top view of one of the globular heads of the neuraminidase, showing the folds of the polypeptide chain, the catalytic cleft at the center, and the positions where amino acid replacements allow the virus to escape neutralization by monoclonal antibodies directed against neuraminidase variant N2. (Reproduced, by permission, from Air et al., 1989) (B) Stereo diagram of β-pleated sheet structure showing hydrogen bonds (dotted). (Reproduced, by permission, from Varghese and Colman, 1991)

Storms, R. W. and H. R. Box, Jr. 1989. Oncogenes, proto-oncogenes and signal transduction: Towards a unified theory. *Adv. Virus Research* **37**:1–34.

Watson, J. D., N. H. Hopkins, J. W. Roberts, J. A. Steitz, and A. M. Weiner 1987. Cancer at the genetic level. In *Molecular Biology of the Gene,* vol. 2. Benjamin/Cummings Pub., Menlo Park, CA, pp. 962–1094.

9
Cytokines, Growth, and Differentiation Factors

Cytokines are small proteins that regulate interactions between cells in the immune system (Arai et al., 1990). Formerly they had to be isolated from biological tissues in trace quantities of doubtful purity, but recombinant DNA technology has made it possible to produce kilograms of pure materials with which to explore their therapeutic potentials. There has been a veritable race to determine their structures by X-ray crystallography and nuclear magnetic resonance. This chapter surveys the structures solved by the end of 1991, but without as yet being able to make much sense of them. Only some of them tell us something about their mechanism of action.

Interleukin-8 and Platelet Factor 4

Interleukin-8 is also known as neutrophil activation factor, monocyte-derived neutrophil chemotactic factor, and T-cell chemotactic factor. As these names indicate, interleukin-8 activates chemotaxis by neutrophils and by a variety of other white blood cells. It is a dimer made up of two identical proteins subunits of 72 amino acid residues each. The atomic coordinates were first determined in solution by magnetic resonance (Clore et al., 1990). Crystallographers then used the coordinates of the monomer to solve the crystal structure after all their usual methods of phase determination had failed—a great triumph for the younger method. (Baldwin et al., 1991).

Figure 9.1 shows superpositions of the 30 most probable structures of the interleukin-8 dimer derived by magnetic resonance. A six-stranded pleated sheet of β-strands supports two α-helices running in opposite directions to form a molecule that looks like the antigen-binding vice of the major histocompatibility protein.

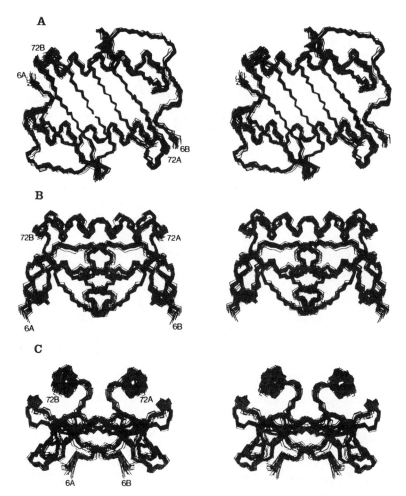

Figure 9.1. Stereoviews of the interleukin-8 dimer in three mutually perpendicular directions. Note the similarity with the α_1 and α_2 domains of HLA-A2 shown in Figure 2.7 on page 53. (Reproduced, by permission, from Clore et al., 1990)

(See Fig. 2.7 on page 53.) In the crystal, the structure of each of the monomers is the same as in solution: after optimal superposition the root mean square difference in individual atomic positions was only 1.1Å. On the other hand, the relative positions of the two monomers in the dimer were different. In solution the angle between the two α-helices was 172° and the perpendicular distance between their centers 14.8Å; in the crystal the angle was 164° and the distance

11.1Å. α-helices have an average girth of 10.5Å, which means that the solution structure leaves a 4.4Å-wide gap between them; in the crystal structure that gap is reduced to 0.6Å, that is, it is virtually closed (Clore and Gronenborn, 1991). In both structures the helices adhere to the underlying pleated sheet with hydrophobic sidechains while they present hydrophilic ones to the surrounding water.

The transition between the two structures is brought about by a small rotation of one monomer relative to the other about an axis that is perpendicular to the twofold symmetry axis relating the two monomers in the dimer, an operation similar to that found in the transition between the alternative quaternary structures in the allosteric enzyme glycogen phosphorylase. In allosteric proteins, different quaternary structures tend to be stabilized by alternative sets of hydrogen bonds and salt bridges between polar groups in the same or adjacent monomers (Perutz, 1989). The same is true here. In interleukin-8 in solution, glutamate 29 of one monomer attracts lysine 23 of its neighbor, while histidine 33 accepts a hydrogen bond from the glutamine 8 of the same monomer. In the crystal, glutamate 4 of one monomer attracts lysine 23 of its neighbor, and histidine 33 donates a hydrogen bond to the backbone carbonyl of glutamate 29 of the same monomer.

Clore and Gronenborn, the authors of the magnetic resonance work, stress that there is as yet no evidence of the closed structure existing in solution, but it seems unlikely that the closing up of the two α-helices is merely an artifact due to crystal lattice forces and bears no relation to the protein's function. On the contrary, I would regard interleukin-8 as a model for other MHC-like proteins such as the 70 kD heat-shock protein, whose activation by ATP suggests an equilibrium between an open and a closed form of the clathrin-binding vice. (See Chapter 2.) Sequence homologies suggest that two other cytokines, the murine growth-related gene product and the macrophage inflammatory protein, have similar mobile vice structures.

Platelet factor 4 is released from activated blood platelets. It attracts other white blood cells such as neutrophils and monocites and may act as a signal that initiates inflammation. Its pharmacological interest lies in its strong binding of the two antithrombotic polysaccharides heparan and heparin. Since heparin also accelerates growth of blood vessels in solid tumors, platelet factor 4 possesses antitumor activity and has been patented under the name oncostatin A. *In vitro* it also binds double-stranded DNA, perhaps because it binds any acid polymer. Platelet factor 4 is a tetramer made up of two interleukin-8–like dimers stuck together back to back, with their vices facing in opposite directions (Fig. 9.2). Each of the helices carries four lysines pointing outward. The 16 lysines could bind a heparin chain wrapped around the molecule, as DNA is wrapped around histones (St. Charles et al., 1989). Alternatively they could aggregate separate heparin molecules.

Interleukin-1β

Interleukin-1β stimulates the proliferation of B-lymphocytes and thymocytes (T-cells) and induces the release of interleukin-2, of prostaglandins, and of collagenase in response to immune reactions and inflammation. It also induces transcription of the cellular oncogene c-Jun. (See page 111.) Its crystal structure has been solved independently by teams in Swiss and American pharmaceutical companies (Finzel et al., 1989; Priestle et al., 1989) and its solution structure was solved by Clore, Wingfield, and Gronenborn at the National Institutes of Health, using an interleukin-1β labeled with ^{13}C and ^{15}N, and expansion of the magnetic resonances into four dimensions. Interleukin-1β is made up of a single chain of 153 amino acid residues, which makes it about half as big again as the largest protein structures previously solved by magnetic resonance, whence the determination of its solution structure represents an important advance (Clore et al., 1991).

The two X-ray and the solution structures are the same within error: the root mean square difference in atomic position between one of the X-ray structures and the solution structure amounts to only 0.85Å for the backbone of the polypeptide chain and 1.3Å for all atoms. The fold of the polypeptide chain is like that of the soya bean trypsin inhibitor (Blow et al., 1974); it is a tetrahedral cage of β-strands filled inside with hydrophobic sidechains and studded with hydrophilic ones on its outside; the β-strands are connected by protruding irregular loops (Figs. 9.3 and 9.4). Investigators have tried to infer its receptor-binding site from the location of external residues that are conserved in different species and from loss of activity caused by genetic replacement of various residues, but the results do not restrict binding to a single site. The tetrahedral structure looks as though it were designed to fit a symmetrically constructed receptor cavity that envelops the interleukin rather than a flat surface or a shallow groove. Alternatively it could serve to draw three receptor molecules together.

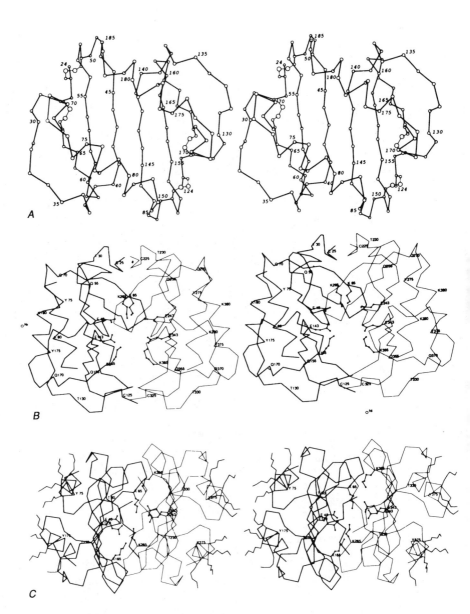

Figure 9.2. Stereoview of platelet factor 4. (A) Dimer showing the characteristic MHC-like vice of two α-helices on a base of β-strands. (B) and (C) Two mutually perpendicular views of the tetramer. In (B) the vices face right and left. In (C) they face up and down. (C) shows the lysine and arginine sidechains that bind heparin. (Reproduced, by permission, from St. Charles et al., 1989)

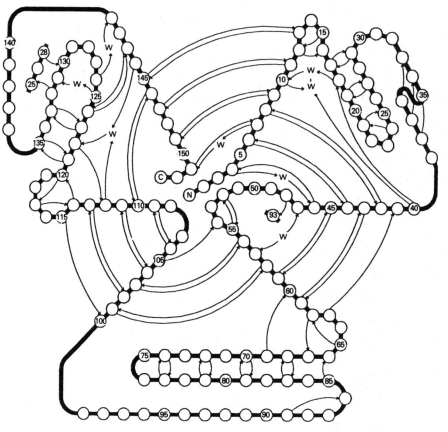

Figure 9.3. Diagram of the β-barrel structure of interleukin-1β. The circles represent α-carbons and the arrows hydrogen bonds from the main chain NHs to main chain COs. (Reproduced, by permission, from Finzel et al., 1989)

Fibroblast Growth Factor

Fibroblast growth factors are a family of which seven members have been identified so far. Zhu and others (1991) and Zhang and others (1991) have determined the crystal structures of two members: the bovine acidic and human basic fibroblast growth factors. The other members are keratinocyte growth factor and four putative oncogene products (FGF 5 and 6, int-2, and hst/KS3). The factors induce cell division and chemotaxis in a variety of cells of skin, nerve, and connective tissue, and also growth of blood vessels. They have a strong affinity for heparin and are sometimes called heparin-binding

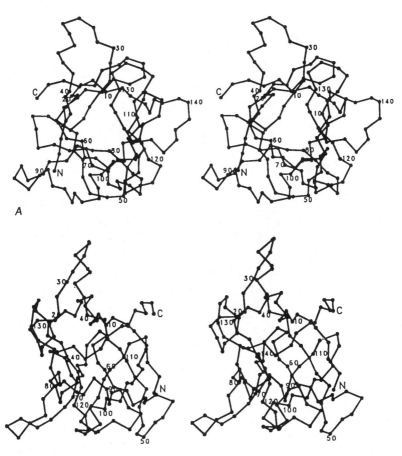

Figure 9.4. Stereoview of interleukin-1β. (A) View along the central axis of the six-stranded β-barrel. (B) View of the molecule turned 90° about the vertical axis in (A). (Reproduced, by permission, from Finzel et al., 1989)

growth factors. Their tertiary structures are similar to that of interleukin-1α and -β (Fig. 9.5). Figure 9.6 shows the probable location of the receptor-binding site and of the seven lysines and arginines that presumably bind heparin.

Interleukin-2

Interleukin-2 is another cytokine that modulates the immune response by interaction with several different cell types. It plays a

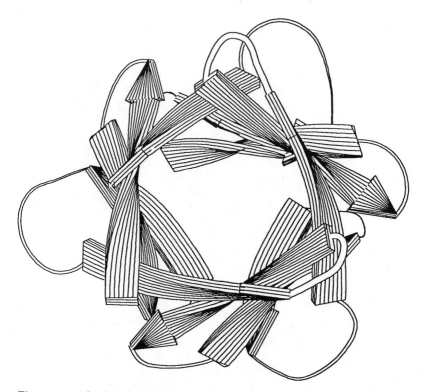

Figure 9.5. Fibroblast growth factor. Schematic view of the interlocked β-strands making up the trimer. (Reproduced, by permission, from Zhu et al., 1991)

Figure 9.6. Fibroblast growth factor. Stereoview of α-carbon positions plus sidechains that interact with the growth factor receptor (bold lines) and basic sidechains that bind heparin (broken lines). (Reproduced, by permission, from Zhu et al., 1991)

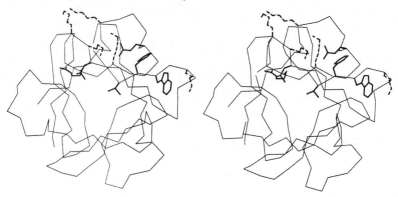

key role in the activity and growth of T-lymphocytes, which synthesize it after they have been stimulated by antigen. Interleukin-2 then triggers the proliferation of T-cells bearing that specific antigen on their receptors. It has an extremely high affinity for the membrane-bound receptor that mediates its activity ($K_D \sim 10^{-11}$); this receptor consists of two or more separate protein molecules (Waldmann, 1989). In contrast to interleukin-1β, interleukin-2 is about 65% α-helical; its structure is unrelated to any other found before (Fig. 9.7) (Brandhuber et al., 1987). Experiments aimed at finding the site or sites of interleukin-2 that bind to the receptor on its target cells include competition between binding of the receptor and binding of different monoclonal antibodies directed at specific sites on the surface of the cytokine; the effects on receptor binding of replacements of single amino acid residues, and the deletion of residues. The results point consistently to parts of helices A, B, and the three carboxy-terminal residues as parts of the receptor-binding site. Knowledge of the interaction of interleukin-2 with its receptor promises to open new avenues to the treatment of leukemia/lymphoma, autoimmune diseases, and rejection of organ transplants (Waldmann, 1991).

Tumor Necrosis Factor

Tumor necrosis factor is a cytokine secreted by macrophages that raised high hopes when it was found to kill human tumor cell

Figure 9.7. Stereodrawing of α-helices, represented as cylinders, and their connections in the structure of interleukin-2. (Reproduced, by permission, from Brandhuber et al., 1987)

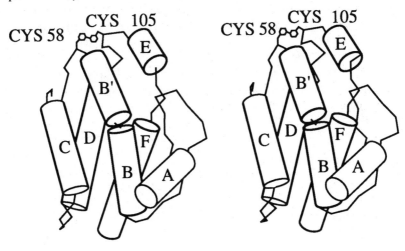

lines derived from breast and colon carcinomas and from a melanoma. It also caused necrosis of human breast carcinoma explants transplanted into nude mice. These hopes were shaken by the discovery that the factor is the same as cachectin, a protein that causes wasting of bones and other tissues and also contributes to the pathogenesis of rheumatoid arthritis and inflammatory tissue destruction (Beutler et al., 1985). It also mediates endotoxin-induced shock, fever, and hypotension, it deranges lipid and glucose metabolism, and it induces acidosis. It has been described as an esential element in many diseases. All the same, there is still great interest in its antitumor activity (Beutler and Cerami, 1988).

Tumor necrosis factor is composed of three identical subunits each containing 157 amino acid residues. Each chain looks more like a β-sandwich than a β-barrel, but the connectivity of its β-strands is the same as that of the viral coat proteins diagrammed in Figure 8.3, on page 189. Figure 9.8 compares the factor to the coat protein of the satellite tobacco necrosis virus that resembles it most closely, except for the outside loops. The trimer is formed by β-strands F and G of one subunit lying across and adhering to the BIDG sheet of the neighboring subunit, mainly by hydrophobic contacts. The arrangement resembles that of the three β-barrels in the hexon of the adenovirus (Figure 9.9.) (Jones et al., 1989; Sprang and Eck, 1989).

Much research has been done to discover the receptor-binding sites of the tumor necrosis factor, but so far there is no clear evidence whether certain amino acid substitutions reduce biological activity because they interfere with receptor binding directly, or because they interfere indirectly by altering the tertiary or quaternary structure. For example, antisera raised against peptides 1–31 and 1–15 blocked receptor binding, but a factor that lacked residues 1–1.27 was still fully active, which suggests that the blocking might have fortuitously covered the different, true binding site. The trimeric structure indicates that the factor may bind to three receptor molecules simultaneously (Sprang and Eck, 1991). For a review of the factor's molecular mechanism of action, see Vilček and Lee (1991).

Interferons and Interleukin 4

The interferons were discovered by experiments on the growth of influenza virus in chick embryos, done primarily for the development of vaccines. Incubation of the embryos with heat-killed virus

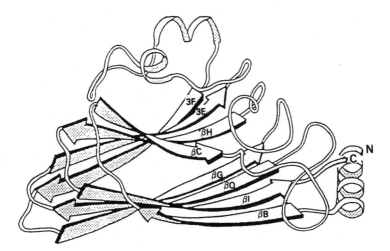

A

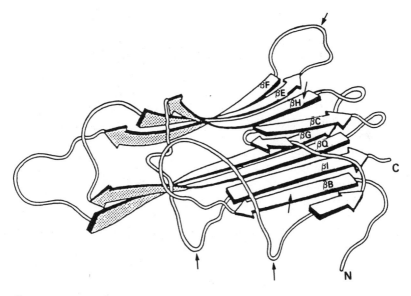

B

Figure 9.8. Comparison of the polypeptide fold in the coat protein of the satellite tobacco necrosis virus (A) and tumor necrosis factor (B). The arrows indicate peptides that are thought to interact with the tumor necrosis factor receptor. The occurence of the word necrosis in the two proteins is coincidental. (Reproduced, by permission, from Jones et al., 1989)

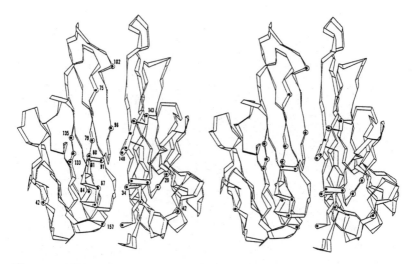

Figure 9.9. Stereoview of two of the three subunits in the trimeric tumor necrosis factor, showing the residues that are the same in human, bovine, and mouse tumor necrosis factors and in the structurally similar lymphotoxin. (Courtesy of Dr. Stephen R. Sprang)

was known to inhibit the growth of live virus. This inhibition used to be attributed to the dead viruses' blocking of the cell surface receptors, but Isaacs and Lindemann in London showed that it was due instead to the release of a soluble factor that interfered with the infection. They called it *interferon* (Isaacs and Lindemann, 1957). Five years later, Paucker, Cantell, and Henle at the University of Pennsylvania discovered that interferon also suppressed the growth of mouse fibroblasts in culture and that the degree of suppression was proportional to the interferon dose. Its activity could be destroyed by trypsin, which proved that it was a protein (Paucker et al., 1962).

At first these discoveries raised high hopes that interferons would prove magic bullets against viruses and cancer, but these hopes could be realized to only a very limited extent as long as only minute quantities of often impure interferons could be extracted from biological tissues. The manufacture of kilograms of pure interferons by recombinant DNA technology has stimulated intense new interest in their structures, mechanism of action, and clinical applications (Petska et al., 1987).

There are three major classes of interferon, α, β, and γ, each giving rise to a multitude of different cytological effects; some of these can be separated as due to different subclasses. One effect common to all classes is the induction of surface expression of class I MHC proteins. Interferon-γ also stimulates surface expression of class II MHC proteins (Petska et al., 1987).

The first structure to be determined was that of murine interferon-β, which is a peptide of 165 amino acid residues folded into five α-helices (Figs. 9.10 and 9.11) (Senda et al., 1990). Strong homologies between the amino acid sequences of murine and human interferon-β and those of interferon-α point to a common structure. This is corroborated by the positions of the disulphide bridges. Human interferon-β contains a bridge connecting cysteine 31 just beyond the N-terminal helix A to cysteine 141 at the beginning of the C-terminal helix E. Human interferon-α contains a bridge

Figure 9.10. Diagram of the fold of the polypeptide chain in murine interferon-β. Arrows give the directions of the α-helices. Numbers give the positions of certain residues. Shaded loops show the probable receptor binding site. (Reproduced, by permission, from Senda et al., 1990)

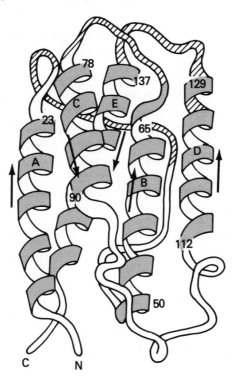

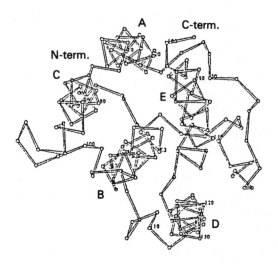

Figure 9.11. α-carbon trace of interferon-β seen from the top of Figure 9.10. (Reproduced, by permission, from Senda et al., 1990)

between cysteine 3 close to the N-terminus and cysteine 97 in the C-D loop. Neither bridge exists in murine interferon-β, but both can be readily constructed in its atomic model, which would not be possible if the three structures differed significantly.

Substitutions of amino acid residues in the loops connecting helices A and B, B and C, or D and E tend to reduce or inhibit the antiviral activity of interferon-β. Peptides with amino acid sequences corresponding to those loops inhibit the neutralizing effect of monoclonal antibodies on its antiviral or antiproliferative action. All these experiments point to the tips of the loops at the top of Figure 9.10 as being the receptor binding sites, just as external loops of small viruses bind to viral receptors.

Human interferon-γ is a dimer composed of two identical peptides of 138 amino acid residues each. Its structure is again largely α-helical (Ealick et al., 1991). As in the dimeric trp-repressor (Zhang et al., 1987), the helices of the two monomers are intimately intertwined and in close contact; helix C is the most buried and also the most hydrophobic of them (Fig. 9.12A). Receptor binding of interferon-γ seems to reside near the N- and C-termini of its two chains, i.e., on the side opposite to the loops that form the receptor site of interferon-β. The dimeric structure of interferon-γ suggests that its cell surface receptor is also a dimer, perhaps with an

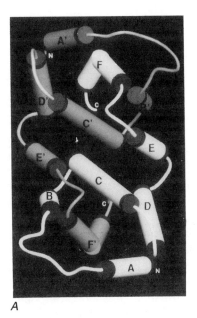

A

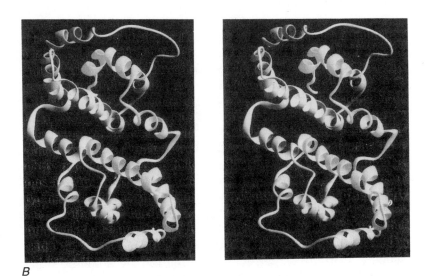

B

Figure 9.12. Structure of interferon-γ. (A) Topology of α-helices in the dimer. (B) Stereoviews of the dimer seen in the same direction as in (A). (Reproduced, by permission, from Ealick et al., 1991)

immunoglobulin-like structure. The topology of the interferon-γ dimer is related to that of the interferon-β monomer in an intriguing way. One half of it resembles the β monomer, but that half is composed of four helices of one chain plus two helices of the other chain (Fig. 9.12B) (Ealick et al., 1991).

α-interferons are now proving effective in the treatment of hairy cell leukemia, chronic myeloid leukemias, and myelomas, even though they are mildly toxic to the bone marrow and liver. γ-interferon is used for the treatment of chronic granulomatous disease (a form of immune deficiency). The antiviral properties of α-interferons have found uses in the treatment of hepatitis B; they are also being tried, together with other drugs, in the treatment of AIDS.

The antiviral and antiproliferative actions of interferons are due to its stimulation of two enzymes: a protein kinase that phosphorylates the eukaryotic initiation factor of protein synthesis eIF2 and thereby inactivates it, and an oligoadenylate synthase that polymerizes ATP to 2′,5′ rather than the usual 3′,5′ polyadenylates. These polyadenylates activate an endonuclease that hydrolyzes messenger and ribosomal RNA. Both activities require the presence of double-stranded RNA (Fig. 9.13) (Farrell et al., 1978). Both inhibit protein synthesis.

Interleukin 4 is a cytokine that stimulates proliferation and differentiation of several types of cells, including B-cells. Prediction of its structure based on its amino acid sequence suggests that it has a 4-helix structure similar to that of the interferon-β (Curtis et al., 1991).

Somatotropin and Its Complex with the Cellular Receptor

The growth hormones are related to the prolactins and to the placental lactogens which are all secreted by the anterior pituitary gland. Growth hormone or somatotroprin is a 22kD protein that has recently found wide medical and commercial uses as a remedy for human dwarfism and retarded growth, as a means of making pigs grow faster and making cows give more milk.

The somatotropin receptor is a protein embedded in the surface membrane of liver cells. On stimulation by somatotropin it transmits a signal into the cell that promotes growth and metabolism of muscle, bone, and cartilage (Nicol et al., 1986). It consists of an extracellular hormone-binding segment, a segment that crosses the membrane and an intracellular segment. Two

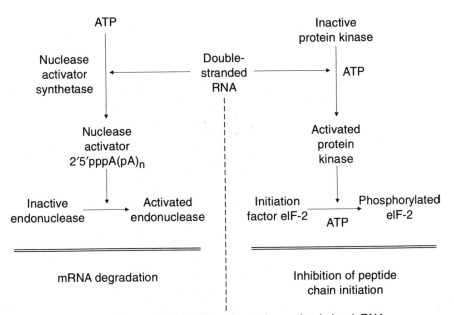

Two pathways for inhibition of protein synthesis by dsRNA

Figure 9.13. Activation of 2,5-polyadenylate synthase and of a protein kinase by interferons and double-stranded RNA. (Reproduced, by permission, from Farrell et al., 1978)

such receptor molecules combine with one molecule of somatotropin (Cunningham et al., 1991). Vos, Ultsch, and Kossiakoff (1992) have published the structure of a complex of somatotropin with two extracellular segments of its receptor. The hormone itself consists of a bundle of four long α-helices connected by long loops that are part helical and part β-strand (Fig. 9.14A). The helices are inclined to each other at angles of about 20°. We have seen in Chapter 3 on page 105 that this angle optimizes the packing of the sidechains between them.

The receptor segment consists of 238 amino acid residues folded to form two immunoglobulin-like domains rigidly connected by a single turn of α-helix and by a salt bridge between an arginine and an aspartate. The N-terminal domain is braced by three disulphide bridges, two more than found in immunoglobulin domains (Fig. 9.14B). In the crystal two receptor segments are bound to each molecule of somatotropin. They are related

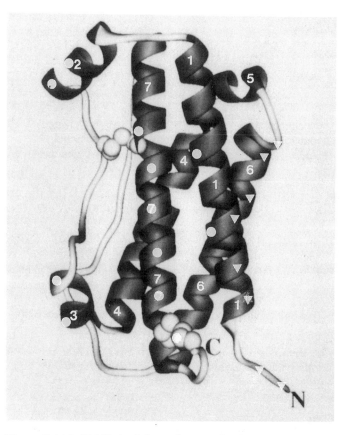

Figure 9.14A. Folding of the polypeptide chain of human somatotropin. The helical segments are numbered 1–7. The white circles indicate contacts with the receptor on the left; the triangles contacts with that on the right. (Courtesy of Dr. B. de Vos and Dr. A. Kossiakoff)

approximately by a twofold axis of symmetry that is vertical in Figure 9.14B. One receptor segment nestles mainly against helices 2, 3, and 7 of the hormone, burying 1230Å^2 of contact area; the symmetry-related segment leans against the hormone's N-terminal segment and helices 1 and 6, burying 900Å^2 of contact area. The larger interface contains nine hydrogen bonds; the smaller one contains four. Combination with the hormone also sticks the receptor's C-terminal domains together, burying another 500Å^2 of contact area and six hydrogen bonds between them.

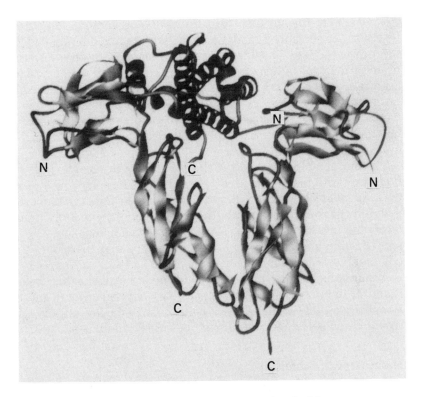

Figure 9.14B. Somatotropin (black helices) sandwiched between two extracellular segments of its receptor. Note the two immunoglobulin-like domains. (Courtesy of Dr. de Vos and Dr. A. Kossiakoff)

The same, largely nonpolar faces of the receptors bind to the two stereochemically distinct faces on either side of the hormone. Most of the receptor's nonpolar sidechains are the same in either contact and have the same conformations even though they are touching different surfaces. There is a good correlation between the residues seen to be in contact in the crystal and those predicted to be in contact by site-directed mutagenesis, which gives one confidence that the structure seen in the crystal is the same as that *in vivo*.

Combination with the hormone joins together those parts of the two receptor segments that would lie closest to the cell membrane in the intact receptor. Mutants of the receptor unable to form dimers on combination with somatotropin are biologically inactive. It seems likely, therefore, that the hormone transmits its signal to the cytoplasm by the very act of joining two receptor molecules together. Conceivably a catalytically active site might be created in the process. I had imagined hormone receptors to be allosteric proteins that transmit signals from one side of a cell membrane to the other by changing their quaternary structure. Instead, the hormone simply makes two separate receptors join and clap their hands. It is also remarkable that nature has evolved a single receptor face that recognizes and is complementary to two stereochemically quite different faces of the hormone.

The amino acid sequence of the somatotropin receptor shows homology with the sequences of other cytokine receptors, including those for interleukines 2, 3, 4, 6, and 7, the granulocyte and granulocyte macrophage colony stimulating factors, and erythropoietin. Tissue factor and interferon receptors show more distant homologies. Di- or trimerization of receptors on hormone binding may therefore be a common way of signal transduction.

A four-helix bundle similar to that of somatotropin has also been found in interleukin-4, which acts as a growth factor for T and B lymphocytes. It also stimulates B cells to make the low affinity IgE receptor and secrete IgE immunoglobulins (Redfield et al., 1991).

Nerve Growth Factor

In the late thirties Guiseppe Levi, professor of anatomy at the University of Turin, and his student Rita Levi-Montalcini (no relation) set out to analyze how excision of as yet non-innervated limbs from early chick embryos affected development of motor cells in the spinal cord and of sensory cells in the dorsal root ganglia. They were led to suspect that the limbs release a tropic factor that is conveyed by the growing axons to their cell bodies in the spinal cord and the dorsal root ganglia. After the war their publication came to the attention of Viktor Hamburger, who had been a pupil of the great German embryologist Hans Spemann. Hamburger invited Levi-Montalcini to spend a semester with him in St. Louis, Missouri. That semester was to become 16 years.

The decisive observations came in an experiment that illustrates the importance of the prepared mind. Elmer Bueker, a former pupil of Hamburger, sent him an article describing how a mouse sarcoma, grafted on to a chick embryo, had become innervated by fibers from the embryo. Bueker concluded that the tumor had provided more ample terrain for the growth of the nerve fibers than the nearby embryonic limb, but Levi-Montalcini thought otherwise. In a euphoric mood, she dropped all current research in order to repeat Bueker's work. On being grafted to her embryos, the tumors Bueker had used became innervated as Bueker had described, but another tumor sent to her (by mistake?) produced a far more dramatic effect. It caused the non-innervated organs of the embryo, including its viscera and blood vessels, to be invaded by large bundles of nerve fibers. She wrote:

> This observation indicated that the tumor had released a humoral, a fluid, factor able both to accelerate differentiative processes in sympathetic and, to a lesser degree, sensory cells; and to cause excessive production, as well as the quantitatively and qualitatively abnormal distribution of nerve fibers. (Levi-Montalcini, 1988)

What was that factor?

Its nature revealed itself by a remarkable stroke of luck, after Stanley Cohen, then a young biochemist, had joined her at St. Louis. In experiments carried out in Carlos Chagas' and Hertha Meyer's laboratory in Rio de Janeiro Levi-Montalcini had established that extracts of mouse tumors induced the formation of halos of nerve fibrils around cultured sensory ganglia, but only if the tumors had first been transplanted into and then excised from chick embryos. Levi-Montalcini and Cohen spent a year trying to extract enough of the factor from such transplanted mouse sarcomas to characterize it, and concluded that it was a nucleoprotein. When Cohen wondered if the nucleic acid might be a contaminant, Arthur Kornberg, then at St. Louis, suggested treating the extract with snake venom whose phosphodiesterase breaks down nucleic acid. But instead of destroying the activity, the snake venom raised it spectacularly, because it contained a several thousand times greater concentration of the suspected growth factor than the mouse tumors. This discovery allowed Cohen to isolate and characterize the factor and provided Levi-Montalcini with pure factor to inject into her embryos. Had they been able to buy pure phosphodiesterase commercially, they would never have discovered this. Later, mouse submaxillary glands also proved a rich source of nerve growth factor.

Early embryos produce a prolific growth of neurons, many of which later die off unless they sense nerve growth factor diffusing from target cells. This mechanism was thought to apply mainly to peripheral, neural crest-derived, sympathetic, and sensory neurons, but similar effects have been detected also in the central nervous system. For example, cholinergic neurons in the basal forebrain respond to the factor. This has aroused great interest, as it may be relevant to Alzheimer's disease. In the forebrain, synthesis of the factor is regulated by neuronal activity. It is raised by the glutamate and lowered by the γ-aminobutyrate transmitter systems (Fig. 9.15) (Thoenen, 1991; Rodriguez-Tébar et al., 1991).

Nerve growth factor is a dimer of two identical polypeptide chains of 118 amino acid residues, each cross-linked by two disulphide bridges. Its structure is even more unusual than that of some of the other cytokines: it is a long bundle of seven β-strands

Figure 9.15. Regulation of nerve growth factor synthesis by nonneuronal cells in the peripheral neurons (A) and by neurons in the central nervous system (B). In the periphery, regulation is independent of input by other neurons. In growth-factor–responsive cholinergic neurons of the basal forebrain, on the other hand, neuronal activity by the glutamate transmitter system raises, and by the γ-aminobutyrate (GABA) transmitter system lowers, the synthesis of growth factor. (Reproduced, by permission, from Thoenen, 1991)

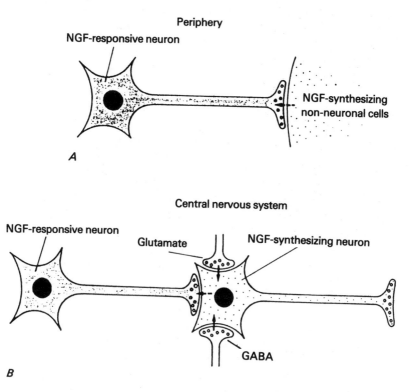

twisted around each other and connected by apparently irregular loops (Fig. 9.16) (McDonald et al., 1991).

Nerve growth factors from several species and related factors all share internal hydrophobic and hydrogen bond-forming residues essential for the stability of the tertiary structure of the individual subunits and for their coherence in the dimer. Most of the variable residues are concentrated in the external hairpin loops 29-35, 43-48, and 92-98 and in three consecutive reverse turns from residues 59-66. It is unclear whether this variability determines the specificity of different factors or is due to random mutation unrestrained by selective pressure. There is evidence suggesting

Figure 9.16. (A) Stereoview of the β-strand structure of the mouse β-nerve growth factor (NGF) subunit. The strands are labelled sequentially starting at the amino terminus. Strands A (residues 15-22) and A''' (residues 40-44) form hydrogen bonds with strand B (residues 47-59). Strand A' (residues 27-30) and A'' (residues 35-38) also form a double-stranded β-sheet that packs in an orthogonal fashion against strands C and D. Strand C(residues 79-93) forms hydrogen bonds with strand D (residues 96-111) to form a double-stranded antiparallel and very twisted β-sheet. β-hairpin turns connect the three pairs of β-strands A' to A'', A''' to B, and C to D. Residues 58-68 contain three consecutive reverse turns. These arestabilized by Pro 63, Gly 67, and Gly 70.

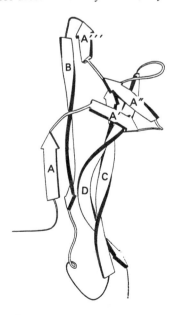

A

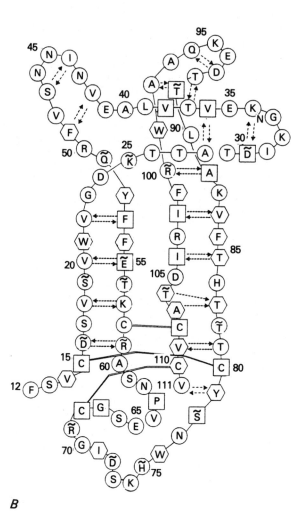

B

Figure 9.16. (B) Schematic representation of the mouse β-NGF subunit highlighting structurally important residues. Bold, absolutely conserved residues for all NGF and NGF-related sequences. Squares, buried residues in the β-NGF subunit with a relative sidechain solvent accessibility of less than 7%. The hexagons represent residues involved in the dimer interface and were identified from an examination of the β-NGF dimer using computer graphics. ~, side chain of the residue participates in a hydrogen bond to either a side-chain or main-chain atom. The main-chain hydrogen bonds involved in the β-sheet structure are displayed as arrows, pointing in the direction of the donor to acceptor. (Reproduced, by permission, from McDonald et al., 1991)

that some residues on strands A and A' interact with the nerve growth factor receptors (Figs. 9.16 and 9.17).

Kinetic studies suggest that there are two kinds of nerve growth factor receptor, one with low and the other with high affinity, but they do not exclude the existence of a single receptor whose affinity rises after the first factor has been bound (Rodriguez-Tébar et al., 1991).

Two other neurotrophic factors with affinities for different target cells have recently been isolated, cloned, and sequenced. They are the ciliary or brain-derived neurotrophic factor and neurotrophin-3. The homology of their amino acid sequences with that of the nerve growth factor testifies to identity of their three-dimensional structures. Directed mutagenesis should soon reveal the chemical basis of their different specificities.

Epidermal Growth Factor

Epidermal growth factor (EGF) regulates growth and replication of cells. It is similar to the transforming growth factor α. Both factors bind to the same receptor on the cell surface that has assumed medical importance since the discovery that mutation of the receptor can be oncogenic. The coat proteins of several viruses include an EGF-like domain with affinity for the EGF-receptor. The course of the polypeptide chain of EGF has been traced by nuclear magnetic resonance. (Campbell et al., 1989; Harvey et al., 1991)

General Remarks

I embarked on this chapter in the hope of arriving at generalizations about the relationship between the structure and function of cytokines, but found tantalizingly few.

Interleukin-8 and platelet factor 4 have been two rewarding structures solved so far, because their striking similarity to the major histocompatibility complex leaves no doubt about the nature of their binding sites, which opens the possibility of synthesizing mutants with modified biological activities or designing antibodies against them. Interleukin-1β, fibroblast growth factor, and tumor necrosis factor are all trimeric; conceivably they all activate cell surface receptors by drawing three of them together. Interleukin-2 and somatotropin, though monomeric, activate receptors by binding two separate ones. Conceivably, kinases might be activated in this way. Muscle hexokinase has a nutcracker structure and is activated

```
          10        20        30        40        50        60
HSDPARRGELSVCDSISEWVTAADKTAVDMSGGTVTVLEKVPVSKGQLKQYFYETKCNP  Pig BDNF
SSSHPIFHRGEFSVCDSVSVWV**GDKTTATDIKGKEVMLGEVNINNSVFKQYFFETKCRD Human NGF
SSSHPVFHRGEFSVCDSISVWV**GDKTTATDIKGKEVMLGEVNINNSVFKQYFFETKCRD Bovine NGF
SSTHPVFHMGEFSVCDSVSVWV**ADKTTATDIKGKEVTVLAEVNVNNNVFKQYFFETKCRD Guinea pig NGF
SSTHPVFHMGEFSVCDSVSVWV**GDKTTATDIKGKEVTVLAEVNINNSVFROYFFETKCRA Mouse NGF
TAHPVLHRGEFSVCDSVSMWV***GDKTTATDIKGKEVTVLGEVNINNNVFKQYFFETKCRD Chick NGF
EDHPVHNLGEHSVCDSVSAWV***TKTTATDIKGNTVTVMENVNLDNKVYKQYFFETCCKN Snake NGF

          70        80        90       100       110
MGYTKEGCRGIDKRHWNSQCRTTQSYVRALTMDSKKRIGWRFIRIDTSCVCTLTIKRGR  Pig BDNF
PNPVDSGCRGIDSKHWNSYCTTTHTFVKALTMDGKO*AAWRFIRIDTACVCVLSRKAVRRA Human NGF
PNPVDSGCRGIDAKHWNSYCTTTHTFVKALTMDGKO*AAWRFIRIDTACVCVLSRKTGQRA Guinea pig NGF
PSPVESGCRGIDSKHWNSYCTTTHTFVKALTTDNKO*AAWRFIRIDTACVCVLNRKAARRG Bovine NGF
SNPVESGCRGIDSKHWNSYCTTTHTFVKALTTDEKQ*AAWRFIRIDTACVCVLSRKATRRG Mouse NGF
PRPVSSGCRGIDAKHWNSYCTTTHTFVKALTMEGKO*AAWRFIRIDTACVCVLSRKSGRP  Chick NGF
PNPEPSGCRGIDSSHWNSYCTETDTFIKALTMEGNQ*ASWRFIRETACVCVITKKKGN  Snake NGF
```

Figure 9.17. Amino acid sequences of nerve growth factors of different species. The shaded regions cover conserved residues.

by the closing of its arms, one binding ATP and the other glucose. One could imagine the two arms being separate receptor molecules which are drawn together by a cytokine. The sparcity of clear relationships between the structures of the cytokines and their functions that have emerged from structural analysis so far is disappointing, but it is still early days. Most of these structures have only just been solved. As I have tried to show elsewhere in this book, the medical value of structural knowledge has often become apparent only much later.

Further Reading

Arai, K.-i, F. Lee, A. Miyajima, S. Miyataka, N. Arai and T. Yakota 1990. Cytokines: Coordinates of immune and inflammatory responses. *Ann. Rev. Biochem.* **59**:783–836.

Beutler, B. and A. Cerami 1988. Tumor necrosis, cachexia, shock and inflammation: A common mediator. *Ann. Rev. Biochem.* **57**:505–18.

Gilman, A. G. 1987. G Proteins: Transducers of receptor-generated signals. *Ann. Rev. Biochem.* **56**:615–50.

Petska, S., J. A. Langer, K. C. Zoon, and C. E. Samuel 1987. Interferons and their actions. *Ann. Rev. Biochem.* **56**:727–78.

Staeheli, P. 1990. Interferon-induced protein and the antiviral state. *Adv. Virus Research* **38**:147–200.

Waldmann, T. A. 1989. The multisubunit interleukin-2 receptor. *Ann. Rev. Biochem.* **58**:875–912.

Yarden, Y. and A. Ulrich 1988. Growth factor receptor protein kinases. *Ann. Rev. Biochem.* **57**:443–78.

10

Some Medically Important Protein Structures Not Discussed in the Preceding Chapters

Protein	Significance
β-lactamases	Penicillins and cephalosporins inhibit bacterial cell wall synthesis. β-lactamases confer resistance to these antibiotics by hydrolizing their β-lactam bond, which is the bond between N and CO in Figure 1.24 on page 29. There are several classes of these enzymes. Class A from *Staphylococcus aureus* and class C from *Citrobacter freundii* have similar structures. A cleft between two domains harbors an active site made up, in class A, of two serines, a lysine, and a glutamate. In class C a tyrosine takes the place of one of the serines. The structure of β-lactamase I from Bacillus cereus is similar to that of one of the penicillin-sensitive enzymes that catalyzes the cross-linking of peptides in the cell wall: D-alanyl-D-alanine carboxypeptidase-transpeptidase from Streptomyces. Knowledge of these structures may help the design of new antibiotics.[1,2,3,4]
Chloramphenicol acetyl transferase	Chloramphenicol binds to the peptidyl transferase center of bacterial ribosomes. Chloramphenicol acetyl transferase makes bacteria resistant to the drug by catalyzing the transfer of an acetyl group to its primary OH.[5]
Thymidylate synthase	This enzyme is essential for DNA synthesis and is therefore a target for anticancer drugs. It catalyzes the reductive methylation of deoxyuridine monophosphate to deoxythymidine monophosphate by CH_2-H_4 folate.[6]

Protein	Significance
Dihydrofolate reductase	This is another enzyme essential for DNA synthesis and a target for anticancer drugs, including methotrexate and trimethoprin.[7,8]
Cyclophilin	This enzyme catalyzes the cis-trans isomerization of prolyl peptides and thereby accelerates the folding of polypeptide chains to their native conformations. It also binds, and is inhibited by, the immunosuppressive drug Cyclosporin, but this may be unrelated to the pharmacological effect of Cyclosporin.[9,37]
Major FK 506 binding protein	This enzyme also catalyzes cis-trans isomerization of proline, but is unrelated to cyclophilin. It binds to, and is inhibited by the immunosuppressive drug FK 506.[10]
Trypanothione reductase	This is the protozoan substitute for mammalian glutathione reductase and therefore a potential target for drugs against trypanosomes and leishmannias. It is a flavoprotein that catalyzes the reduction of the bis(glutathione) spermidine-conjugate trypanothione.[11]
Surface proteins of *Trypanosoma brucei*	Trypanosomes rapidly change their protein coats in response to immunological pressure. There may be a thousand different surface proteins with variable sequences having the same fold. An understanding of the folding pattern may offer new approaches to drug design.[12]
Glycosomal triose phosphate isomerase from *Trypanosoma brucei*	Part of a concentrated effort to determine the structure of trypanosomal enzymes that might lead to the development of drugs inhibiting the trypanosomal but not the equivalent human enzymes.[13]
Acetylcholinesterase	Enzyme in synapses of cholinergic neurons which hydrolyzes the neurotransmitter acetylcholine. Its active site lies at the end of a 20Å deep gorge lined with aromatic residues. Ester hydrolysis is catalyzed by a triad of a glutamate, a histidine, and an aspartate.[14]
Neurophysin II	Neurophysins I and II are disulfide-rich proteins bound, respectively, to oxytocin and vasopressin. They help to pack high concentrations of these hormones into neurosecretory granules in the anterior pituitary.[15]
Human pancreatic lipase	This enzyme hydrolyzes lipid triglycerides first into di- and then into monoglycerides and free fatty acids. This is the first essential step in fat absorption.[16]

Protein	Significance
Rheumatoid arthritis fluid phospholipase A_2	Phospholipase A_2 hydrolyzes the ester bond between C_2 of glycerol and the fatty acid attached to it. It is associated with inflammation and tissue damage because it releases arachidonate, a precursor of prostaglandins.[17,18,19]
Cholera toxin-related enterotoxin and verotoxin-1 from *E. coli*	Members of a family of toxins that includes cholera and shiga toxins. The cholera-related enterotoxin has a fascinating structure which shows how the toxin works. It is made up of two different subunits, A and B. The A-subunit catalyzes the activation of adenylate cyclase in the affected cell, thereby causing high concentrations of cyclic adenosine-3',5',-monophosphate that expels fluid and ions from the cell. The five B-subunits form a pore that is filled by the C-terminal peptide of the A-subunit. The pentamer has a flat surface that attaches itself to the phospholipids at the surface of the target cell and facilitates the entry of the catalytic subunit into the cell. The structure of the B-subunit of verotoxin-1 is similar, even though its sequence shows no homology to that of the cholera-related enterotoxin.[20,35,36]
Exotoxin of *Pseudomonas aeroginosa* and diptheria toxin	In these two toxins the functions of receptor binding, membrane translocation, and catalysis are contained in separate domains along a single chain. In the exotoxin the N-terminal receptor domain has a β-barrel structure like the viral coat proteins. The central membrane translocation domain is α-helical like other membrane proteins, and the C-terminal catalytic domain is made of eight β-strands and seven α-helices and has a nutcracker structure like the kinases. The closely similar catalytic domain of diptheria toxin is N-terminal; its central membrane translocation domain is made of nine antiparallel α-helices; its C-terminal domain is again a β-barrel.[21,22]
Catalytic subunit of cyclic adenosine-3',5'-monophosphate–dependent protein kinase	Attachment of certain hormones to cell surface receptors activates catalytic subunits on the cytoplasmic face of the receptors. These subunits synthesize cyclic adenosine-3',5'-monophosphate (c-AMP), which then acts as a second messenger, phosphorylating, and thereby activating, other proteins. The c-AMP–dependent kinase catalyzes that phosphorylation.[23]

Protein	Significance
Ubiquitin	Ubiquitin forms peptide bonds between its own C-terminal carboxylate and the amino groups of lysines of cellular proteins, thereby initiating their ATP-dependent degradation when they are either misfolded or denatured.[24]
Transthyretin (formerly called prealbumin)	Transthyretin is one of two thyroxin-binding proteins in blood plasma. It crosses the blood-brain barrier and is the only thyroxin-carrying protein in the cerebrospinal fluid. It is made of four identical peptides of 127 residues folded into two layers of β-strands and grouped around a hydrophobic, 50 Å long and 8 Å wide channel that contains two deeply buried thyroxin binding sites.[25,26]
Retinol binding protein	Carrier for retinol or vitamin A that circulates in plasma in association with transthyretin. Its two-layers β-barrel encloses a largely hydrophobic cavity tailored to fit retinol. Retinol binding protein is a member of a large family of ligand-transport proteins of similar structure that includes β-lactoglobulin and bilin binding protein.[27,28]
β-lactoglobulin	Protein in the whey of milk that binds retinol, though it is not certain that this is its physiological function. Structurally homologous to retinol binding protein, but the retinol binding site is in a surface pocket rather than buried in the β-barrel.[29,30]
Human lactoferrin (or lactotransferrin)	Lactoferrin binds iron with a dissociation constant of about 10^{-20}, but releases it to specific receptors *in vivo*. It occurs in milk and other excretions and is also secreted by leukocytes. Its high affinity for iron gives it a strong bacteriostatic action. Its polypeptide chain of about 700 residues is folded into two similar lobes related by near two-fold symmetry, and each lobe has two domains that chelate one atom of ferric iron between them. The structure of lactoferrin is similar to those of plasma transferrin, ovotransferrin and also to a family of bacterial periplasmic proteins specific for binding L-arabinose, D-galactose, leucine/isoleucine/valine, sulfate and others (for review see reference 32). Binding or release of substrate in these proteins may be accomplished by the bending of two domains about a hinge.[31]

Protein	Significance
Human serum albumin	Serum albumin is the most abundant plasma protein. It binds a wide range of ligands, including metals, lipids, amino acids, hormones, and many drugs. Its single chain of 585 amino acid residues is composed of three genetically duplicated domains, each of which is again divided into two similar motifs. The structure of serum albumin reflects this sequence homology (Fig. 10.1). Each motif is made up of similarly folded helices, some of which have been shown to harbor binding sites for aspirin and valium. It will now become possible to modify the affinities of drugs for serum albumin on a rational basis.[33]
Muscle glycogen phosphorylase	Glycogen phosphorylase is the key enzyme in the mobilization of chemical energy from glycogen. It is a complex allosteric protein that is activated by phosphorylation of one of its serine OHs and by AMP, and inhibited by glucose-6-phosphate. Its stereochemical mechanism has been clarified by X-ray analysis of its active phosphorylated and its inactive dephosphorylated structures.[34]
Ferritin	Ferritin is the principal iron storage protein. It contains about 4500 Fe^{3+} ions inside a spherical shell of 24 identical, symmetrically disposed protein subunits containing over 160 residues in a single chain. That chain is folded into four antiparallel helices. There are two types of human ferritin, L and H. H-subunits contain a two-iron catalytic center that may catalyze the oxidation from ferrous to ferric iron necessary for storage. [39]
Adenosine deaminase	Adenosine deaminase catalyzes the hydrolysis of the amino group from adenosine and deoxyadenosine. Its absence or malfunction causes severe combined immune deficiency disease. Its 363-residue-long chain is folded into a barrel lined with eight β-strands inside and eight α-helices outside, similar to triose phosphate isomerase and at least 16 other enzymes. Its catalytic site contains a zinc ion tetrahedrally coordinated to three histidines and an aspartate. Many of its pathogenic mutations cause amino acid substitutions affecting, either directly or indirectly, that site.[40]

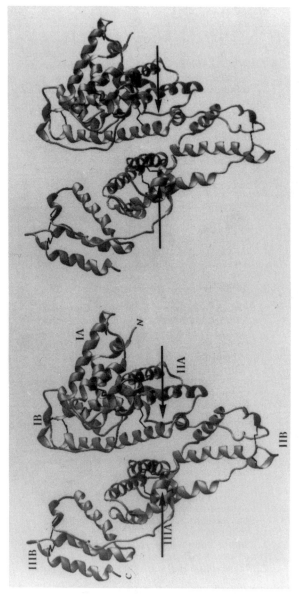

Figure 10.1. Stereodrawing of the tertiary structure of human serum albumin, showing the homologous domains I, II, and III, each divided into two similar subdomains A and B. Each subdomain is made up of α-helices in a globin-like fold, linked by cystine bridges; these are shown as zigzag bars (⌐_⌐). The C-terminal domain IIIB is most clearly resolved, with its two cystine bridges visible right at the top. Another cystine bridge shows up well at the bottom of the figure in IIB. The two arrows point to drug binding sites. Warfarin bound to IIA (right arrow); ibuprofen, AZT, and digitoxin to IIIA (left arrow), and aspirin to both sides. (Courtesy of Dr. Daniel Carter)

References

1. Herzberg, O. 1991. Refined crystal structure of β-lactamase from *Staphylococcus aureus* PC1 at 2.0Å resolution. *J. Mol. Biol.* **217:**701–19.
2. Oefner, C., A. d'Arcy, J. J. Daly, K. Gubernator, R. L. Charnas, I. Heinz, C. Hubschwerlen, and F. K. Winkler 1990. Refined crystal structure of β-lactamase from *Citrobacter freundii* indicates a mechanism of β-lactam hydrolysis. *Nature* **343:**284–88.
3. Knox, J. R., and P. C. Moews 1991. β-Lactamase of *Bacillus licheniformis* 749/C. Refinement at 2Å resolution and analysis of hydration. *J. Mol. Biol.* **220:**435–55.
4. Sutton, B. J., R. J. Todd, P. J. Artymiuk, S. G. Waley, and D. C. Phillips 1986. Tertiary structural similarity between a class A β-lactamase and a penicillin-sensitive D-alanyl carboxypeptidase-transpeptidase. *Nature* **320:**378–80.
5. Leslie, A. G. W. 1990. Refined crystal structure of type III chloramphenicol acetyltransferase at 1.75Å resolution. *J. Mol. Biol.* **213:**167–86.
6. Montfort, W. R., K. M. Perry, E. B. Fauman, J. S. Finer-Moore, G. F. Maley, L. Hardy, F. Maley, and R. M. Stroud 1990. Structure, multiple site binding, and segmental accommodation in thymidylate synthase on binding dUMP and an anti-folate. *Biochemistry* **29:**6964–77.
7. Matthews, D. A., R. A. Alden, J. T. Bolin, S. T. Freer, R. Hamlin, N. Xuong, J. Kraut, N. Williams, M. Poe, and K. Hoogsten 1977. Dihydrofolate reductase: X-ray structure of the binary complex with methotrexate. *Science* **197:**452–55.
8. Oefner, C., A. d'Arcy, and F. K. Winkler 1988. Crystal structure of human dihydrofolate reductase complexed with folate. *Eur. J. Biochem.* **174:**377–86.
9. Kallen, J., C. Spitzfaden, M. G. M. Zweini, G. Wider, H. Widmer, K. Wüthrich, and M. D. Walkinshaw 1991. Structure of human cyclophilin and its binding site for cyclosporin A determined by X-ray crystallography and NMR spectroscopy. *Nature* **353:**276–79.
10. Moore, J. M., D. A. Peattie, M. J. Fitzgibbon, and J. A. Thomson 1991. Solution structure of the major binding protein for the immunosuppressant FK506. *Nature* **351:** 248–50.

11. Kuriyan, J., X. -P. Kong, T. S. R. Krishna, R. M. Sweet, N. J. Murgolo, H. Field, A. Cerami, and G. B. Henderson 1991. X-ray structure of trypanothione reductase from *Crithidia fasciculata* at 2.4Å resolution. *Proc. Nat. Acad. Sci. USA* **88**:8764–68.

12. Metcalf, P., M. Blum, D. Freymann, M. Turner, and D. C. Wiley 1987. Two variant surface glycoproteins of *Trypanosoma brucei* of different sequence classes have similar 6Å resolution X-ray structures. *Nature* **325**:84–86.

13. Wierenga, R. K., K. H. Kalk, and W. G. Hol 1987. Structure determination of the glycosomal triosephosphate isomerase from *Trypanosoma brucei brucei* at 2.4Å resolution. *J. Mol. Biol.* **198**:109–21.

14. Sussman, J., M. Harel, F. Frolow, C. Oefner, A. Goldman, L. Toker, and I. Silman 1991. Atomic structure of acetylcholinesterase from *Torpedo californica:* A prototypic acetylcholinesterase-binding protein. *Science* **253**:872–79.

15. Chen, L., J. P. Rose, E. Breslow, D. Yang, W. R. Chang, W. F. Furey, Jr., M. Sax, and B. C. Wang 1991. Crystal structure of a bovine neurophysin II dipeptide complex at 2.8Å determined from the single-wavelength anomalous scattering signal of an incorporated iodine atom. *Proc. Nat. Acad. Sci. USA* **88**:4240–44.

16. Winkler, F. K., A. d'Arcy, and W. Hunziger 1990. Structure of human pancreaic lipase. *Nature* **343**:771–74.

17. Thunnissen, M., A. Eiso, K. H. Kalk, J. Drenth, B. W. Dijkstra, O. P. Kuipers, R. Dijkman, G. H. de Haas, and H. M. Verheij 1990. X-ray structure of phospholipase A2 complexed with a substrate-derived inhibitor. *Nature* **347**:689–91.

18. Wery, N. P., R. W. Schevitz, D. K. Clawson, J. L. Bobbitt, E. R. Dow, G. Gamboa, T. Goodson, Jr., R. B. Hermann, R. M. Kramer, D. B. McClure, E. D. Mihelich, J. E. Putnam, J. D. Sharp, D. H. Stark, C. Teater, M. W. Warrick, and N. D. Jones 1991. Structure of recombinant human rheumatoid arthritic synovial fluid phospholipase A2 at 2.2Å resolution. *Nature* **352**:79–82.

19. Scott, D. L., S. P. White, Z. Otwinowski, W. Yuan, M. H. Gelb, P. B. Sigler, 1990. Interfacial catalysis: the mechanism of phospholipase A2. *Science* **250**:1541–46.

20. Sixma, T. K., S. E. Pronk, K. H. Kalk, E. S. Wartna, B. van Zanten, B. Witholt, and W. G. J. Hol 1991. Crystal structure of a cholera toxin-related heat labile enterotoxin from *E. coli. Nature* **351**:371–77.

21. Choe, S., M. J. Bennett, G. Fujii, P. M. J. Curmi, K. A. Kantardjieff, R. J. Collier and D. Eisenberg 1992. Three domains for three functions: The crystal structure of diphtheria toxin. *Nature,* in press.
22. Allured, V. S., R. J. Collier, S. F. Carroll, and D. B. McKay 1986. Structure of exotoxin A of *Pseudomonas aeruginosa* at 3.0Å resolution. *Proc. Nat. Acad. Sci. USA* **83**:1320–24.
23. Knighton, D. R., J. Zheng, L. F. Ten Eyck, V. A. Ashford, N. H. Xuong, S. S. Taylor, and J. M. Sowardski 1991. Crystal structure of the catalytic subunit of cyclic adenosine monophosphate-dependent protein kinase. *Science* **253**:407–13.
24. Vijay-Kumar, S., C. E. Bugg, K. D. Wilkinson, and W. J. Cook 1985. 3-D Structure of ubiquitin at 2.8Å resolution. *Proc. Nat. Acad. Sci. USA* **82**:3582.
25. Blake, C. C. F., M. J. Geisow, S. J. Oatley, B. Rerat, and C. Rerat 1978. Structure of prealbumin: Secondary, tertiary and quaternary interactions determined by Fourier refinement at 1.8Å. *J. Mol. Biol.* **121**:339–56.
26. Blake, C. C. F., and S. J. Oatley (1977) Protein-DNA and protein hormone interactions in prealbumin: A model of the thyroid hormone nuclear receptor. *Nature* **268**:115–20.
27. Newcomer, M. E., T. A. Jones, J. Aquist, J. Sundelin, U. Eriksson, I. Raski, and P. A. Petersen 1984. The three-dimensional structure of retinol-binding protein. *EMBO J.* **3**:1451–54.
28. Cowan, S., M. E. Newcomer, and T. A. Jones 1990. Crystallographic refinement of human serum retinol binding protein at 2Å resolution. *Proteins: Structure, Function & Genetics* **8**:44–61.
29. Papiz, M. Z., L. Sawyer, E. E. Eliopoulos, A. C. T. North, J. B. C. Findlay, R. Sivaprasadarao, A. T. Jones, M. E. Newcomer, and P. J. Kraulis 1986. The structure of β-lactoglobulin and its similarity to plasma retinol binding protein. *Nature* **324**:383–85.
30. Monaco, H. L., G. Zanotti, P. Spadon, M. Bolognesi, L. Sawyer, and E. E. Eliopoulos 1987. Crystal structure of a trigonal form of bovine β-lactoglobulin and of its complex with retinol at 2.5Å resolution. *J. Mol. Biol.* **197**:695–706.
31. Anderson, B. F., H. M. Baker, E. J. Dodson, G. E. Harris, S. V. Rumball, J. M. Waters, and E. M. Baker 1987. Structure of lactoferrin at 3.2Å resolution. *Proc. Nat. Acad. Sci. USA* **84**:1769–73.

32. Quicho, F. A. 1991. Atomic structures and functions of periplasmic receptors for active transport and chemotaxis. *Current Opinion in Structural Biology*, Vol. 1, 1922–33.

33. He, X.-m., Y. Satow, Z. Otwinowski, E. Casale, P. D. Twigg, and D. C. Carter 1992. Structure and chemistry of human serum albumin at 3.2Å resolution. *Nature,* submitted.

34. Barford, D., S. -H. Hu, and L. N. Johnson 1991. Structural mechanism for glycogen phosphorylase control by phosphorylation and AMP. *J. Mol. Biol.* **218:**233–60. For review see Perutz, 1989.

35. Stein, P. E., A. Boodhoo, G. J. Tyrell, J. L. Brunton, and R. J. Read 1992. Crystal structure of the cell-binding B-oligomer of verotoxin-1 from *E. coli. Nature* **355:**748–50.

36. Sixman, T. K., S. E. Pronk, K. H. Kalk, B. A. M. van Zanten, A. M. Berghuis and W. G. J. Hol 1992. Lactose binding to heat-labile enterotoxin revealed by X-ray crystallography. *Nature* **355:**561–64.

37. De Franco, A. L. 1990. Immunosuppressants at work. *Nature* **352:**754–55.

38. Banyard, S. H., D. K. Stammers and P. M. Harrison 1978. Electron density map of ferritin at 2.8Å resolution. *Nature* **271:**282–84.

39. Lawson, D. M., P. J. Artymiuk, S. J. Yewdall, J. M. A. Smith, J. C. Livingstone, A. Treffry, A. Luzzago, S. Levi, P. Arosio, G. Cesareni, C. D. Thomas, W. V. Shaw and P. M. Harrison 1991. Solving the structure of human H ferritin by genetically engineering intermolecular crystal contacts. *Nature* **349:**541–44.

40. Wilson, D. K., F. B. Rudolph, F. A. Quiocho 1991. Atomic structure of adenosine deaminase complexed with a transition-state analog: understanding catalysis and immunodeficiency mutations. *Science* **252:**1278–84.

11

Benefits to Medicine, Past, Present, and Future

The living cell is like a symphony orchestra without a conductor: its score is laid down in its DNA, and proteins are its performing instruments. Scientists began studying their structures because they wondered what they looked like and how they worked, without expecting their results to bear on human disease. Bernal and Crowfoot could not have guessed in 1934 that their discovery of X-ray diffraction by pepsin crystals was to be the first step in the search for a drug against AIDS, a disease then unknown, and in the rational design of antihypertensive drugs, nor could they have had any inkling of nature's intriguing design and mechanism of action that solution of its structure would one day reveal.

In 1937 I chose X-ray analysis of hemoglobin for my Ph.D. thesis because the structure of proteins seemed the most fundamental problem in biochemistry, and hemoglobin was easiest to crystallize. Hemoglobin diseases were not yet known to exist, so it could not have occurred to me that solution of its structure might one day clarify their molecular pathology, let alone enable people to engineer a hemoglobin as a blood substitute.

In 1973 Roberto Poljak's solution of the first immunoglobulin structure, a mouse Fab fragment, immediately revealed the stereochemical basis of immunological specificity (Poljak et al, 1973), but at the time the mere concept of using that for the design of genetically engineered antibodies would have been relegated to science fiction. The recent prolongation of the lives of two terminally ill leukemia patients bears witness to the power and promise of combined structural analysis and gene technology in medicine. As of 1991, the structures of the human histocompatibility complex and of the T-cell receptor CD4 are still too new for medical applications to have materialized, but there is little doubt that they will come.

Development of antihypertensive and antiviral drugs has shown how much faster and cheaper it can be to design drugs when the atomic structures of their receptors are known, instead of having to be inferred by trial and error. In this field, the main barrier to future advance lies in the difficulty of crystallizing membrane proteins. So far, the structures of only six such proteins have been solved: the photosynthetic reaction center (Deisenhofer et al., 1985), the purple membrane protein bacteriorhodopsin (Henderson et al., 1990), porin of photosynthetic bacteria (Weiss et al., 1990), a plant light harvesting complex (Kühlbrandt and Wang, 1991) and two fungal peptides that form ion channels in phospholipid membranes: gramicidin A and zervamicin (Wallace and Ravikumar, 1988; Langs et al., 1991; Karle et al., 1991); at first sight none of them seem to be pharmacologically relevant, except by way of illustrating the general principles of membrane protein architecture, but this view may soon be proved wrong.

Bacteriorhodopsin is made up of a bundle of seven membrane-spanning α-helices connected by peptide loops. Attached to one of the helices and buried between them lies the photoreceptor retinal (Fig. 11.1). Bacteriorhodopsin acts as a light-driven proton pump that generates chemical energy by creating an electrical potential gradient across the bacterium's phospholipid membrane.

In the 1980s biochemists determined the amino acid sequences of many eukaryotic membrane receptors. Among them is a family of receptors whose membrane-spanning domains may be structurally related to that of bacteriorhodopsin, because their sequences indicate that they are also made up of seven transmembrane α-helices. They include the muscarinic acetylcholine, the L-dopamin, the serotonin, the angiotensin II, and many neuropeptide receptors. No chemical homology has been found between the amino acid sequences of their helices and those of bacteriorhodopsin, but their sequences do offer hints of structural homology; if confirmed, then the structure of the membrane protein from a photosynthetic bacterium in a salt lake in California may become medically relevant. N. Unwin and R. Henderson began their attack on bacteriorhodopsin in 1973 for the sole reason that it was the only membrane protein then available in crystalline form, albeit only as two-dimensional crystals. Its possible wider relevance did not become apparent until 1990.

Structural analysis itself has benefitted greatly from gene technology, not only because proteins available in nature in only minute concentrations can now be produced in quantity in *E. coli,*

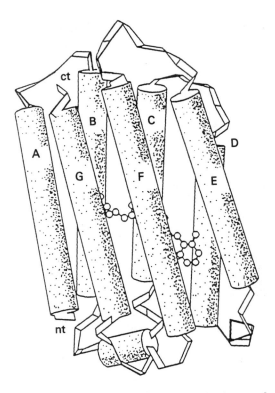

Figure 11.1. Diagram of the structure of bacteriorhodopsin showing the arrangement of its seven transmembrane α-helices and of its buried retinal. (Reproduced, by permission, from Henderson et al., 1990)

yeast, or cultures of mammalian cells, but also because one of protein crystallography's hardest steps, the preparation of isomorphous heavy atom derivatives, has been eased. Nagai and others have solved the structure of the RNA-binding domain of a small ribonuclear protein that is part of the complex involved in the excision of introns from messenger RNA. They first synthesized the protein in *E. coli* and then used gene technology to replace each of its glutamines in turn by cysteines. Reaction of each of the cysteine mutants with mercurials provided the crystalline, isomorphous heavy atom derivatives needed to solve the structure (Nagai et al., 1990). Alternatively, selenocysteines can be incorporated into proteins in place of cysteines or other amino acid residues, and anomalous X-ray scattering by the selenium atoms can be used to determine the phases (Yang et al., 1990). (See Chapter 1.)

Since the early eighties we have also seen spectacular technical and theoretical advances in X-ray crystallography, electron diffraction, and nuclear magnetic resonance spectroscopy that have allowed the structures of ever larger molecules to be determined faster and with greater precision.

Each new protein structure has been a revelation full of surprises. Their medical significance could rarely have been guessed until they had been solved, and medical applications often had to wait until other, as yet unforeseen new techniques had been developed.

It would therefore be rash to make any specific predictions about future progress in the medical applications of protein structures, but if the past 10 years are any guide to the next, there is little doubt that progress will accelerate.

Further Reading

Henderson, R., J. M. Baldwin, T. A. Ceska, F. Zemlin, E. Beckmann, and K. H. Downing 1990. Model for the structure of bacteriorhodopsin based on high resolution electron cryomicroscopy. *J. Mol. Biol.* **213**:899–929.

Kühlbrandt, W. 1988. Three-dimensional crystallization of a membrane protein. *Quart. Rev. Biophys.* **21**:429–77.

Appendix 1

Mathematical Principles of X-ray Analysis

By a theorem due to the French mathematician Joseph Fourier, any periodic function $\psi(x)$ may be represented by the sum of an infinite series of sine and cosine terms. If the function has period c and if $\alpha = 2\pi(x/c)$, then

$$\psi(x) = \frac{a_0}{2} + a_1 \cos \alpha + a_2 \cos 2\alpha + a_3 \cos 3\alpha + \ldots$$
$$+ b_1 \sin \alpha + b_2 \sin 2\alpha + b_3 \sin 3\alpha + \ldots \tag{A1.1}$$

Cos α, cos 2α, cos 3α, etc. represent waves of length c, $c/2$, $c/3$, which are symmetrical about the origin at $x = 0$. Sin α, sin 2α, sin 3α, etc. represent waves of the same length that are antisymmetrical about the origin. a_0 is a constant; a_1, a_2, a_3, etc. and b_1, b_2, b_3, etc. are the amplitudes of the waves that form the *coefficients* of the Fourier terms.

Fourier's theorem implies that any periodic function can be decomposed into a fundamental and a series of harmonics. Using the properties of trigonometric functions, we can write down the formulae for the coefficients in Equation A1.1, namely:

$$a_n = \frac{2}{c} \int_0^c \psi(x) \cos \frac{2\pi n x}{c} \, dx \tag{A1.2}$$

$$b_n = \frac{2}{c} \int_0^c \psi(x) \sin \frac{2\pi n x}{c} \, dx \tag{A1.3}$$

We call this Fourier analysis because from Equation A1.3 we can compute the strengths of the waves of different frequency contained in $\psi(x)$. (See Fig. A1.1.) We call the coefficients in the series *the Fourier transform of the function.*

The amplitudes of the fundamental and harmonics of the transform are given by $|F_1| = \sqrt{a_1^2 + b_1^2}$, etc. Their phase angles are given by

$\tan \varphi_1 = \dfrac{b_1}{a_1}$, etc. Each term may be represented by a radius vector of length $|F|$ that makes an angle φ with the horizontal base, just like the amplitudes and phases of the X-ray reflexions in the Argand diagram of Fig. 1.17 on page 23. The straight brackets $|F|$ mean "modulus," i.e., the magnitude of F.

In 1915, only two years after W. L. Bragg had solved the first crystal structures, his father W. H. Bragg pointed out that the atomic distribution in a crystal could be represented by a Fourier series, and that the amplitudes and phases of its terms would correspond to those of the different orders of diffraction from that crystal. He realized, in fact, that the very process of diffraction consists in a Fourier analysis of the structure (Bragg, 1915).

Suppose we project all the electron density ρ of the atomic structure onto a line normal to a set of lattice planes of spacing a. Then the amplitude and phase of the first order diffraction is given by the sum

$$F_1 = \frac{1}{a} \left\{ \rho(x_1) \cos 2\pi \left(\frac{x_1}{a}\right) + \rho(x_2) \cos 2\pi \left(\frac{x_2}{a}\right) + \ldots \right.$$
$$\left. \rho(x_1) \sin 2\pi \left(\frac{x_1}{a}\right) + \rho(x_2) \sin 2\pi \left(\frac{x_2}{a}\right) + \ldots \right\} \qquad \text{(A1.4)}$$

for all values of ρ between $x = 0$ and $x = a$. For example, if we measure ρ at a hundred intervals between $x = 0$ and $x = a$, then we have to sum 100 sine and 100 cosine terms for the fundamental and for each of the harmonics. Note that all atoms contribute to every term in the series, except at inflection points where the sine or cosine terms pass through zero.

The same analysis can be applied to microscopic vision by substituting for the electron density in the crystal the fraction of light transmitted by the object as coefficients of the series. The summation will then give the amplitudes and phases of the optical diffraction pattern, for instance, those of the Spectra S_1, S_2, S_3 diffracted by the grating in Fig. 1.3 on page 5.

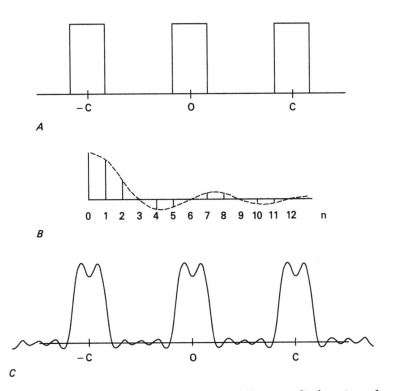

Figure A1.1. Fourier analysis and synthesis. (A) A periodic function, of period c, consisting of rectangular apertures. (B) The Fourier transform of the function in (A), plotted as a spectrum showing the strengths a_n of the various waves of different frequency. The symmetry of the particular function in (A) implies that, in equation A1.2, $b_n=0$ and that $a_n=0$ when n is a multiple of 3. (C) Resynthesis of the function in (A) using just the terms up to n = 5 from the spectrum. Because the series has been truncated, omitting the higher-frequency terms, this curve is a blurred or low-resolution version of the starting curve in (A). The curve in (C) could be made to look more and more like (A) by including higher and higher frequency terms in the series.

If we move the apertures in diagram A further and further apart, then the spectra in B move closer and closer together. When the separation between the apertures becomes infinite, the spectra merge into the continuum outlined by the broken curve. This curve represents the Fourier transform of a single aperture. The spectra in B sample that transform at intervals of $n\lambda/c$ where n is the order of the spectra and λ the wavelength. (Courtesy of Dr. R. A. Crowther)

To reconstitute either the optical or the X-ray image we have to perform a Fourier synthesis. The amplitudes S at any point of x of the image of a grating with lines drawn at intervals of a are given by the sum

$$S(x_1) = C + A_1 \cos 2\pi \left(\frac{x_1}{a}\right) + A_2 \cos 2\pi \left(\frac{2x_1}{a}\right) + A_3 \cos 2\pi \left(\frac{3x_1}{a}\right) + \ldots$$

$$+ B_1 \sin 2\pi \left(\frac{x_1}{a}\right) + B_2 \sin 2\pi \left(\frac{2x_1}{a}\right) + B_3 \sin 2\pi \left(\frac{3x_1}{a}\right) + \ldots$$

$$(A1.5)$$

where C is a constant giving the total amplitude of light transmitted, $A_1^2 + B_1^2 = F_1^2$, the intensity of the first order spectrum, and $\dfrac{B_1}{A_1} = \tan \alpha_1$, its phase. Just as every point in the grating contributed to each of the diffracted spectra, so each of the spectra contributes to every point of the image.

To obtain the electron density distribution normal to a set of lattice planes from the X-ray diffraction spectra we substitute $\rho(x)$ for $S(x)$, the total number of electrons Z for C, and call F_1, the amplitude of the first order X-ray diffraction, etc.

Suppose we wish to obtain a two-dimensional image of the crystal structure. Then we must calculate the contribution of each of many hundred or thousand X-ray reflexions to the point $x = 1$, $y = 1$ in the projection of the unit cell on a plane with the help of the function

$$\rho(x_1y_1) = \frac{1}{ab \sin \gamma} \left\{ Z + A_{11} \sin 2\pi \left(\frac{x_1}{a} + \frac{y_1}{b}\right) + \right.$$

$$A_{21} \sin 2\pi \left(\frac{2x_1}{a} + \frac{y_1}{b}\right) + A_{12} \sin 2\pi \left(\frac{x_1}{a} + \frac{2y_1}{b}\right) + \ldots$$

$$B_{11} \cos 2\pi \left(\frac{x_1}{a} + \frac{y_1}{b}\right) + B_{21} \cos 2\pi \left(\frac{2x_1}{a} + \frac{y_1}{b}\right) +$$

$$\left. B_{12} \cos 2\pi \left(\frac{x_1}{a} + \frac{2y_1}{b}\right) + \ldots \right\},$$

$$(A1.6)$$

and similarly for all other points x,y in a grid covering the plane. γ is the angle between the a and b axes of the unit cell. Similarly with terms $\left(\dfrac{x_1}{a} + \dfrac{y_1}{b} + \dfrac{z_1}{c}\right)$ in three dimensions. The terms would be the mathematical equivalents of the fringes drawn in Figure 1.23 on page 28.

If the structure is centrosymmetric, like hexamethylbenzene or phthalocyanin considered earlier, then the sine terms vanish, because for every term at x where $\sin 2\pi(x/a)$ is positive, there is an equivalent term at $-x$ where $\sin 2\pi(x/a)$ is negative. In centrosymmetric structures, therefore, the electron density at the point $x = y = z = 1$ would be

$$\rho(x_1 y_1 z_1) = \frac{1}{V}\left\{ Z_0 + F_{111} \cos 2\pi \left(\frac{x_1}{a} + \frac{y_1}{b} + \frac{z_1}{c}\right) + \right.$$

$$F_{211} \cos 2\pi \left(2\frac{x_1}{a} + \frac{y_1}{b} + \frac{z_1}{c}\right) +$$

$$\left. F_{221} \cos 2\pi \left(2\frac{x_1}{a} + 2\frac{y_1}{b} + \frac{z_1}{c}\right) + \ldots\right\} \tag{A1.7}$$

the sum being calculated for all the reflexions. V is the volume of the unit cell.

Textbooks of X-ray crystallography rarely write out the individual terms of the Fourier series, but define it by a triple summation. For centrosymmetric structures

$$\rho(xyz) = \frac{1}{V}\left\{ \sum_h \sum_k \sum_l F_{hkl} \cos 2\pi \left(h\frac{x}{a} + k\frac{y}{b} + l\frac{z}{c}\right)\right\}. \tag{A1.8}$$

This means sum the contributions of every X-ray reflection for every point in the units cell with coordinates x,y,z. For noncentrosymmetric crystals, rather than writing out the separate triple sums of the sine and cosine terms, mathematicians define vectors that have both a magnitude and a phase as complex numbers that are the sum of a real and an imaginary term, $\mathbf{F} = A + iB$. The expression for the electron density then becomes an exponential

$$\rho(xyz) = \frac{1}{V}\sum_h \sum_k \sum_l |F_{hkl}| \; e^{i\alpha_{hkl}} e^{2\pi i\left(h\frac{x}{a} + k\frac{y}{b} + l\frac{z}{c}\right)} \tag{A1.9}$$

where $|F|$ is the modulus of $\mathbf{F}$.

The overwhelming part of the X-ray diffraction pattern arises from scattering by the electrons rather than the atomic nuclei, whence atoms cannot be treated as point scatterers, but must be treated as spheres of gradually diminishing electron density. Unlike the pinholes used in our optical analogue which contribute a uniform positive amplitude to the entire diffraction pattern, interference between the X-rays scattered by electrons at different radii from the atomic nucleus reduces the amplitude of the scattered X-rays with increasing angle θ from the incident beam. Figure A1.2A shows that reduction for several light atoms, calculated from electron distributions that had been derived by wave mechanics. Calculation of the amplitude and sign of an X-ray reflection from a crystal of hexamethylbenzene therefore requires each term in the Fourier series to be given a coefficient equal to the total number of electrons in the unit cell multiplied by the fraction of electrons whose contribution is not extinguished by interference at that particular angle of diffraction θ.

Another factor that must be taken into account is thermal motion. Atoms vibrate, which makes them appear larger to the X-rays than they really are. Therefore their scattering amplitude falls off even faster with angle of diffraction, as shown by the broken line in Figure A1.2A. To allow for the thermal motion, the atomic scattering factors are multiplied by a Gaussian function. If $\bar{\mu}^2$ is the mean

Figure A1.2. (A) Full lines: amplitudes of X-rays of wavelength λ scattered by atoms of oxygen, nitrogen, and carbon, at rest, in numbers of electrons, as a function of the angle θ with the incident beam. Broken line: scattering by a carbon atom with a vibration of mean amplitude of 0.1Å. (B) Amplitudes of X-rays scattered by chlorine atoms and chlorine ions. These curves are known as atomic form factors. They represent Fourier transforms of the electron density distributions of these atoms calculated from wave mechanics. (Reproduced, by permission, from Dunitz, 1979)

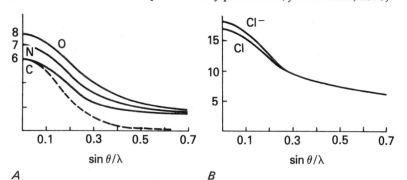

amplitude of vibration and $B = 8\pi^2\bar{\mu}^2$, then the amplitude scattered at the angle θ by the atom at rest is multiplied by the exponential $e^{-B(\sin \theta/\lambda)^2}$.

So much for the basic principles. A real X-ray analysis of a protein structure starts by recording the intensities of the reflections from tens or hundreds of thousands of different lattice planes. Each reflection is given three indices (h, k, and l) according to the corresponding point in the reciprocal lattice. Their phases may then be determined by isomorphous replacement with heavy atoms, combined with anomalous scattering by these atoms. The accuracy of the phase angle of each reflection can be gauged by determining the distribution of probability of the phase angle around the circle of an Argand diagram. If the circles from all the isomorphous replacements intersect at a point as in Figure 1.18 on page 24, there is no problem, but if they don't, the centroid of the probability function is taken as the best amplitude and phase, as in Figure A1.3. This introduces a weighting factor that gives unambiguous terms their full amplitudes, but reduces the amplitudes of ambiguous ones in proportion to the departure of the probability from unity.

Blow and Crick (1959) and Dickerson et al. (1961) have shown the mean square error of the density in a centroid Fourier to be

$$<\Delta\rho>^2 = \frac{2}{V^2} \sum_h \sum_k \sum_l F_{hkl} (1-m^2_{hkl}), \qquad (A1.10)$$

where m is the "figure of merit" of the reflexion hkl, defined as the weighted mean of the cosine of the error in phase angle α

$$m_{hkl} = \frac{\displaystyle\int_0^{2\pi} p(\alpha) \cos (\alpha-\alpha_0) \, d\alpha}{\displaystyle\int_0^{2\pi} p(\alpha) \, d\alpha} = <\cos \Delta\alpha>. \qquad (A1.11)$$

The position of the most probable phase is taken as α_0. The authors show that $m = |F_{best}| / |F|$, i.e., it represents the ratio of the weighted structure amplitude used in the centroid Fourier over the observed amplitude.

Figure A1.4 is a plot of the distribution of figures of merit for all the 1200 reflections used as coefficients in the first three-

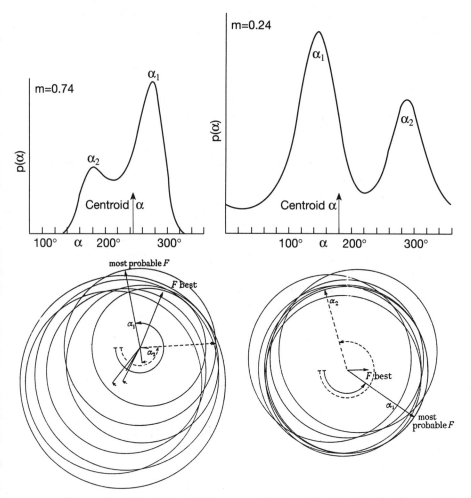

*Figure A1.3. (Top) Argand diagrams showing the phase circles for two of the reflections from hemoglobin crystals where multiple isomorphous replacement gave ambiguous phase angles, because the circles crossed at several points, rather than at a single one as in Fig. 1.18. (Bottom) The curves show the probabilities of α plotted around the circle. Each has two maxima. The arrows show the magnitude and phase of the centroid of the probability distribution. These were the values of the structure amplitudes **F** used for calculating the electron density map of hemoglobin at 5.5Å resolution. (Reproduced, by permission, from Cullis et al., 1961)*

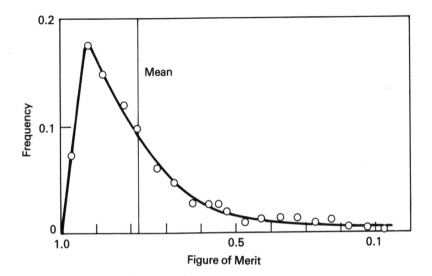

Figure A1.4. Distribution of figures of merit m of phase angles plotted against frequency of reflections used to calculate the electron density distribution in hemoglobin at 5.5Å resolution. (Reproduced, by permission, from Cullis et al., 1961)

dimensional Fourier synthesis of hemoglobin at 5.5Å resolution. It has a peak at 0.92 and a mean value of 0.78; according to Equation (A1.11) these values correspond to standard errors in the phase angle of 23° and 37°, respectively. The electron density distribution is now calculated with the help of a three-dimensional Fourier synthesis. The standard error in electron density, calculated according to Equation (A1.10), from the figures of merit for all reflections is 0.12 electrons/Å^3.

If the electron density map gives a clear image of the structure, then an atomic model of the protein can be constructed. The accuracy of the atomic positions can be gauged by subjecting the model to a Fourier analysis and calculating the intensities of the X-ray reflections. The reliability factors

$$R = \frac{\sum_h \sum_k \sum_l \left(F_0 - F_c \right)}{\sum_h \sum_k \sum_l F_0} \quad \text{or} \quad \frac{\sum_h \sum_k \sum_l \left(F_0 - F_c \right)^2}{\sum_h \sum_k \sum_l F_0^2} \tag{A1.12}$$

give a quantitative measure of the accuracy. F_o stands for the observed and F_c for the calculated amplitudes.

Initially the R-factor may be large. It can be reduced by refining the structures with the help of computer algorithms that shift atomic positions methodically to minimize the sum

$$\sum_h \sum_k \sum_l \left(F_0 - F_c\right)^2.$$ After one such cycle of refinement the atomic

positions are shifted by another, different cycle so as to reduce the free energy due to departures of bond distances, bond angles, hydrogen bonds, and van der Waals contacts from their standard values. These two cycles are repeated in turn until no further improvement in R occurs. Before fast computers made such refinement practicable, it was difficult to improve the accuracy of protein structures to R-factors of much less than 0.3, but modern methods allow them to be cut down to between 0.2 and 0.15. Nevertheless this is still many times larger than the R-factors that can be achieved in the X-ray analysis of small organic compounds, because the ratio of observable X-ray reflections to atomic parameters is much smaller for proteins.

Modern methods of refinement have become so powerful as to be capable of pushing the atoms even of partially wrong structures into positions that yield favorable R-factors. Brünger has pointed out that such self-deception can be avoided by omitting a fraction, say 10%, of randomly chosen reflections from the refinement cycles. Their R-factors are consistently worse for partially wrong structures than for correct ones (Brünger, 1992).

For nearly 20 years after W. H. Bragg first proposed the application of Fourier series to X-ray analysis, series with the intensities rather than the amplitudes as coefficient were believed to have no physical meaning, but in 1934 the Canadian crystallographer A. L. Patterson showed that this was untrue. A function P of the form

$$P(x,y,z) = \frac{1}{V^2} \sum_h \sum_k \sum_l \left\{ F_{hkl}^2 \cos 2\pi \left(h\frac{x}{a} + k\frac{y}{b} + l\frac{z}{c} \right) \right\} \qquad (A1.13)$$

gives the distribution of interatomic vectors radiating from a common origin (Fig. A1.5) (Patterson, 1934). In other branches of physics, this is known as an autocorrelation function, but in X-ray analysis it became known as "the Patterson." It was used with great success to solve moderately complex structures such as amino acids and

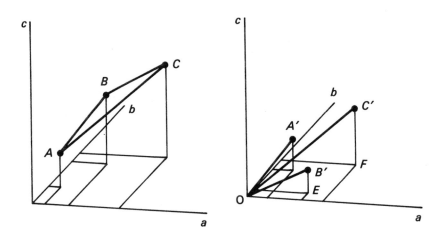

Figure A1.5. Left: atoms in unit cell. Right: vectors from the origin O;
$A' = A{\to}B$; $B' = B{\to}C$; $C' = A{\to}C$. *(Reproduced, by permission, from*
Bunn, 1945)

small carbohydrates, but since the number of interatomic vectors is
$n(n{-}1)$, where n is the number of atoms in the unit cell, it did not solve
the structure of any protein. In 1949, a three-dimensional Patterson
of hemoglobin that I had obtained after several years of laboriously
collecting and densitometering, by eye, thousands of reflections on
hundreds of X-ray oscillation photographs, proved uninterpretable.

In 1953 I found that if $|F|$ is the amplitude of a reflection of the
parent protein and $|F_H|$ that of its heavy atom derivative, then a
Fourier series, calculated with $(|F_H|{-}|F|)^2$ as coefficients, gave
the vectors between the two mercury atoms attached to the reactive
sulphydryl groups of hemoglobin (Fig. A1.6). I called this a difference Patterson (Green et al., 1954). Some years later, Kendrew and
I were faced with the problem of determining the relative positions
of the heavy atoms in different heavy atom derivatives, when M. G.
Rossmann came to join us and showed that the problem could be
solved by taking $(|F_{H1}|{-}|F_{H2}|)^2$ as coefficients of a Fourier series,
where $|F_{H1}|$ and $|F_{H2}|$ are the amplitudes of reflections from the
two different heavy atom derivatives. Here is his argument:

Let the vectors $\mathbf{F}(OP)$, $\mathbf{F}_1(OQ)$, $\mathbf{F}_2(OR)$ be the representations on the
Argand diagram of the structure factors of the plane (hkl) for the un-
substituted compound, and for the isomorphous heavy atom derivatives

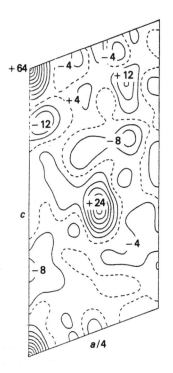

Figure A1.6. Difference Patterson of mercury versus native hemoglobin showing a peak of density 24 compared with a density of 64 at the origin. The peak marks the end of a vector from the origin giving the distance and direction of separation of the two mercury atoms bound to the reactive cysteines of hemoglobin. (Reproduced, by permission, from Green, Ingram, and Perutz, 1954)

1 and 2 respectively [Fig. A1.7]. Thus if $\mathbf{f}_1$ and $\mathbf{f}_2$ are the heavy-atom contributions alone in compounds 1 and 2, then $\mathbf{F}_1 = \mathbf{F} + \mathbf{f}_1$ and $\mathbf{F}_2 = \mathbf{F} + \mathbf{f}_2$. Let QR be the vector $\mathbf{f}$, let $\angle ROQ = \varphi$ and let $\angle RPQ = \psi$.

It will be shown that a Patterson synthesis with $(|F_1| - |F_2|)^2$ as coefficients is approximately equivalent to the self-Patterson of the heavy atoms in compound 1, plus the self-Patterson of the heavy atoms in compound 2, minus the cross-Patterson between the heavy atoms in compounds 1 and 2. That is to say, the Patterson will give positive peaks at the end of vectors between atoms in the same compound. but negative peaks at the end of the vectors between atoms in the two different compounds [Fig. A1.8]. In $\triangle ROQ$, $\mathbf{f}^2 = F_1^2 + F_2^2 - 2F_1F_2 \cos \varphi$.

The difference in phase, φ, between the phase angles of the heavy-atom derivatives 1 and 2 will be small if the heavy-atom contribution in each

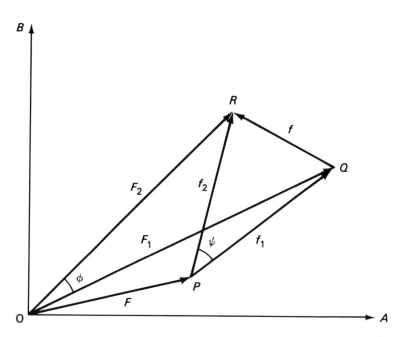

Figure A1.7. Vector diagram showing the relationship between structure factors F in different isomorphous heavy atom derivatives. For explanation see text. (Reproduced, by permission, from Rossmann, 1960)

of the compounds is only a small fraction of the total scattering material, as must inevitably occur in the majority of reflections in a protein structure. Thus to a first order approximation $\cos \varphi = 1$. Hence

$$f^2 \cong F_1^2 + F_2^2 - 2F_1F_2 \cong (F_1 - F_2)^2 \qquad (A1.14)$$

where F_1 and F_2 are the magnitudes of the measured structure factors. (Rossmann, 1960)

Thus, even though Rossmann's difference Patterson was an approximation, it solved the problem in hand. Without knowledge of the relative positions of the heavy atoms in the different derivatives, the structures of myoglobin and hemoglobin could not have been solved.

In the 1950s several crystallographers realized that the phases of X-ray reflections from crystals cannot be entirely random because organic structures are made of spherical atoms of roughly equal size and weight, and the density between them can never be negative.

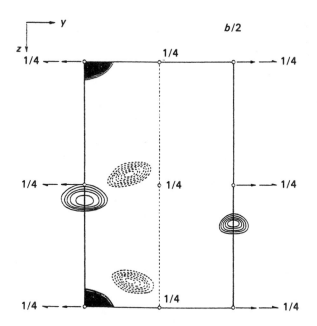

Figure A1.8. Composite view of several sections through the three-dimensional difference Pattersons of two different mercury derivatives of hemoglobin. The two positive peaks along the vertical axis lie at the ends of vectors separating mercury atoms in the same derivative. The negative peaks, drawn with broken lines, lie at the ends of vectors separating mercury atoms in the two different derivatives. (Reproduced, by permission, from Rossmann, 1960)

Their study of phase relations culminated in computer algorithms that solve the structures of moderately complex compounds directly from merely the measured intensities of the X-ray reflections. Such direct methods have also been used to determine the heavy atom positions in the RNA-binding protein U1A after difference Patterson maps had proved uninterpretable (Nagai et al., 1990). Several mathematicians are now trying to develop direct methods also for the determination of protein structures themselves, and G. Sheldrick at the University of Göttingen has succeeded in determining the structure of the small iron-sulphur protein rubredoxin and of the pancreatic polypeptide in this way, but he has pointed out to me that these were very special cases, because the X-ray reflections extended far beyond the usual range, to spacings of about 1Å, and because the FeS_4 complex in rubredoxin and the zinc ion in the polypeptide were helpful, even though he did not use them directly (Sheldrick, 1990).

In highly symmetric molecules the multiplicity of structurally identical protein subunits imposes restrictions on the phase angles which have proved powerful aids in structure determination, especially of viruses (Bricogne, 1976; Harrison et al., 1978).

Further Reading

Dunitz, J. D. 1979. *X-ray Analysis and the Structure of Organic Molecules.* Cornell University Press, Ithaca, NY.

Lipson, H. and W. Cochran 1966. *The Determination of Crystal Structures.* 3rd edition. G. Bell & Sons, Oxford, England. A lucid introduction for those trained in physics.

Appendix **2**

Principles of Protein Structure

The Amino Acids

Proteins consist of amino acids linked together by peptide bonds. The commonly occurring amino acids are of 20 different kinds which all contain the same dipolar ion group $H_3\overset{+}{N}\cdot CH\cdot COO^-$ (Fig. A2.1). The -NH·CH·CO-group, derived from it by the elimination of H_2O, forms the backbone of the polypeptide chain (Fig. A2.2). Specificity is provided by the 20 different kinds of sidechains R attached to the α-carbon.

By convention, the carbon atoms of the amino acids are designated by Greek letters, the CH group that forms part of the main chain being called α, the first carbon of the side chain β, and so on. The α-carbon carries four different chemical substituents in all amino acids except glycine. Of the two possible enantiomorphs, only the L form occurs in proteins. Its configuration is:

Taking the hydrogen atom as the apex of a tetrahedron and looking

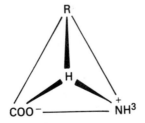

down on it, the sidechain R, and the amino and carboxyl groups succeed each other in a clockwise direction.

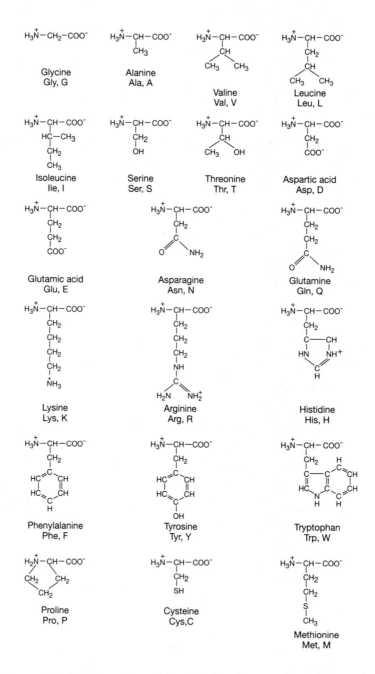

Figure A2.1. The 20 amino acids, with three-letter and one-letter codes. (Reproduced, by permission, from Perutz, 1962)

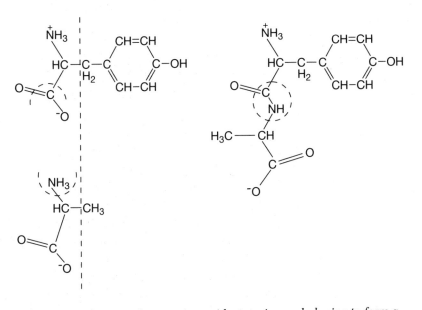

Figure A2.2. Linkage of two amino acids, tyrosine and alanine to form a peptide bond (in circle). This leaves an amino group at the top and a carboxyl group at the bottom free to form peptide bonds with further amino acids. In this way a long polypeptide chain can be formed. (Reproduced, by permission, from Perutz, 1962)

The simplest amino acid is glycine where R is a hydrogen atom (Fig. A2.1). Next come alanine, valine, leucine, and isoleucine, with nonpolar sidechains of increasing length, which may act as spacers in the interior of protein molecules. Serine and threonine carry aliphatic hydroxyl groups capable of forming hydrogen bonds with suitable donor or acceptor groups, such as the imino nitrogen or the carbonyl oxygen of the main polypeptide chain. Serine also acts as a carrier of, and reagent with, phosphate to which it attaches itself by an ester bond, and it forms part of the catalytic site of many hydrolytic enzymes and of the lining of ion channels.

The sidechains of glutamic and aspartic acid both carry carboxylic groups that remain ionized throughout the physiological pH range. Their amides asparagine and glutamine can act as both donors and acceptors of hydrogen bonds and can thus contribute to the internal coherence of protein molecules or to their solubility in water. The carboxyl groups of glutamic and aspartic acids often act as proton donors or acceptors in catalysis.

Serines, threonines, and asparagines are also the binding sites of carbohydrates that are attached to the surfaces of many proteins. Carbohydrates bound to serines and threonines form O-glycosidic bonds and those linked to asparagines form N-glycosidic bonds.

Lysine and arginine carry cationic groups that are also ionized at physiological pH. They and the two carboxylic acids are the principal contributors to the electric charge of protein molecules and give them their properties of dipolar ions. Minor contributors are the terminal carboxyl and amino groups of the polypeptide chain, which are also ionized, and the histidines. These occupy a unique position because they can change their state of ionization within the physiological range of pH. Their imidazole sidechain carries a positive charge, provided by an extra proton, below about pH 6.3 (the exact pK_a varies with the environment) and is neutral above this pH. For this reason, and because it forms coordination complexes with iron and other metals, histidine is often found at the catalytic site of enzymes. Of the three amino acids with aromatic sidechains, tyrosine and tryptophan carry hydrogen-bonding groups. Phenylalanine is nonpolar.

The δ-carbon atom of the sidechain of proline is linked to the imino nitrogen of the main chain, thereby inhibiting its participation in hydrogen bond formation and making proline a misfit in any helical polypeptide chain, except in its three N-terminal positions.

Cysteine carries the highly reactive suphydryl group. This does not ionize at physiological pH nor form hydrogen bonds of significant strength, but two cysteines placed some distance apart along a polypeptide chain, or forming part of different chains, can be joined by oxidation to form the disulphide bridge of cystine which plays an important part in stabilizing protein structures. In one enzyme, cytochrome *c*, cysteines form links between the protein and the heme. Cysteines also bind zinc, copper, and iron ions. The sulphur atom in methionine is unreactive and generally serves no function other than imposing a special configuration on the aliphatic sidechain, but in cytochrome *c* it forms the link between the protein and the heme iron.

The stereochemical configuration of the amino acids, their interatomic distances and bond angles, and the absolute configuration of the L and D forms are accurately known from X-ray analyses of their crystal structures.

The Primary Structure of Proteins

An enzyme molecule may consist of one or more polypeptide chains that, together, may contain between a hundred and several thousand amino acid residues, all arranged in a definite, genetically determined sequence. The longest chain yet discovered is titin, with over 27,000 residues. It is a muscle protein that spans half the length of a sarcomere (Labeit et al., 1990). The number of chains and the sequence of residues within them constitute the primary structure of proteins. By convention, the residues are numbered in sequence along each chain, beginning from the amino end. If there are several chains in one molecule they are designated by Roman or Greek letters which are added to the residue numbers. Analysis of the primary structure of proteins presented insuperable difficulties until the development of chromatographic methods. These formed the experimental basis for Sanger's attack on the constitution of insulin which reached its climax in 1955 with the complete elucidation of the sequence of its 51 residues in two closely linked chains (Fig. 7.1 on page 174). Sanger's discovery was one of the milestones in protein chemistry. First of all, it removed the last shadow of doubt from the polypeptide hypothesis enunciated by Hofmeister more than 50 years earlier. It established the fact that the amino acid residues really are arranged in a definite, genetically determined sequence, but disproved the widely held belief that this sequence was regular. It revealed the part played by cystine bridges in the architecture of protein molecules and the chemical nature of species specificity. Most important of all, Sanger demonstrated that the complete formula of a protein can be determined by chemical methods, at least as far as the primary covalent bonds are concerned, and he thereby stimulated a great volume of new research all over the world (Ryle et al., 1955).

Many enzymes contain one or more nonprotein (prosthetic) groups which form the sites of their catalytic activity. These may be metal ions or metal-organic compounds, as in the respiratory carriers and in certain respiratory enzymes, or pigments related to certain B vitamins, or nucleic acid derivatives, or others. It is a characteristic feature of these enzymes that the prosthetic groups by themselves do not possess the catalytic activity, and that this is conferred on them only by combination with a *specific* protein. Combination with different proteins may confer different catalytic activities on the same prosthetic group, like the heme in catalase and peroxidase.

The Configuration of the Polypeptide Chain:
Secondary Structure

Long-chain polymers in which the same atomic pattern repeats at regular intervals tend to have screw symmetry. This means that each unit of pattern, a methylene or an isoprene group, say, is brought into congruence with its neighbors along the chain by a rotation about a common axis and a translation along it. Polypeptide chains are no exception, despite the diversity of their sidechains. The nature of the sidechains does, however, decide the most stable among several possible screw symmetries.

At first sight it would appear that a polypeptide chain might be able to assume a very large number of different configurations, but in fact rotation is restricted to a greater or lesser extent about each of the three different bonds making up the chain, and no configuration is stable unless it allows every imino group to be hydrogen-bonded to a carbonyl belonging either to the same chain or to a neighboring one. The exact nature of these stereochemical restrictions was first pointed out by Pauling, Corey, and Branson (1951).

Structures with Hydrogen Bonds Between Chains
Three structures of this type are known. The simplest is a planar zigzag chain, which occurs in synthetic polyamides (nylon). Neighboring chains are joined to form sheets by hydrogen bonds between carbonyl and imino groups, while neighboring sheets cohere through the residual forces provided by nonpolar contacts. Pauling and Corey (1951) discovered that sidechains could not be accommodated if the polypeptide chains in the hydrogen-bonded sheets were fully extended, and that this difficulty could be overcome by pleating the sheets at right angles to the chain direction. This has the effect of making the $C\alpha$–$C\beta$ bond extend at right angles to the plane of the sheet and allows sidechains to be neatly packed. In the pleated sheet each residue is related to its neighbors along the chain by a rotation of 180° and a translation of between 3.25 and 3.5Å (Fig. A2.3). Pleated β-sheets normally have a righthanded twist that alleviates close contacts between the β-carbon atoms of one residue and the carbonyl oxygen atoms of the next residue along the chain that would exist in a flat sheet. A left-handed twist would aggravate those contacts (Fig. A2.4) (Chothia et al., 1977). Cowan and McGavin (1955) discovered another type of structure in which residues are related by a rotation of 120° and a translation of 3.1Å (Fig. A2.5). This occurs as a left-handed helix in the synthetic

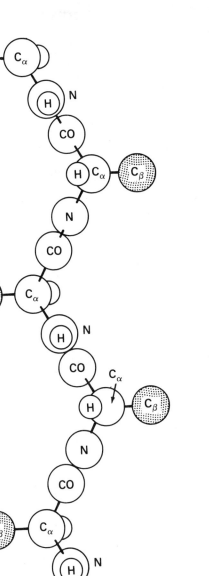

Figure A2.3. (A) Pleated β-strand; alternate sidechains protrude on opposite sides. If one side of a β-strand faces the solvent, and the other side faces inside a protein, sidechains therefore tend to be alternatively hydrophilic and hydrophobic.

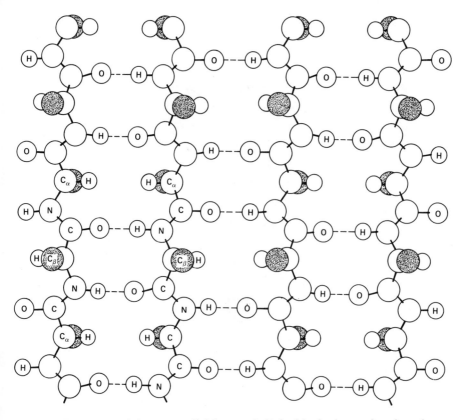

Figure A2.3 (B) Antiparallel β-strands linked by hydrogen bonds to form a pleated sheet. The pleats are normal to the plane of the paper. Alternate sidechains protrude from this sheet in opposite directions. Similar sheets made of parallel β-strands are also found, and so are mixed parallel and antiparallel sheets. (Adapted from L. Pauling, The Nature of the Chemical Bond, *3rd ed., Cornell University Press, Ithaca, NY, 1960)*

polypeptides poly-L-proline and poly-L-hydroxyproline and in one modification of polyglycine (Crick and Rich, 1955a). It is important because it forms the structural basis of collagen (Crick and Rich, 1955b).

Structures with Hydrogen Bonds Within the Same chain
All structures of this type are helical. Sets of atoms can form hydrogen-bonded rings. Depending on whether any imino group is joined to a carbonyl group in one or the other direction along the chain, two families of helices can be constructed. Formally, any

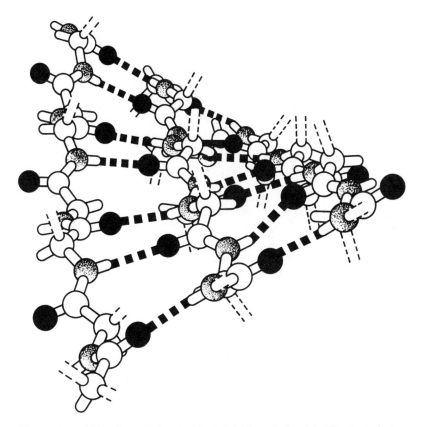

Figure A2.4. A β-pleated sheet with a right-handed twist. The first chain is antiparallel to the second, which is parallel to the third. (Courtesy of Dr. Cyrus Chothia, 1977)

particular structure can be characterized by three symbols: S, N, and r or l. S denotes the number of residues per turn of the helix; N is the number of atoms in the hydrogen-bonded ring, which must be either $3n+4$ or $3n+5$, n being an integer; r or l stands for right- or left-handed.

Donohue (1953) examined all possible structures of this type and calculated their strain energies due to deviations from optimal bond angles and short van der Waals contacts. He arrived at the conclusion that the α-helix of Pauling, Corey, and Branson (1951) was the only form free from strain energy (Fig. A2.6). According to the preceding nomenclature it is designated as $S_N=3.6_{13}$. Three

286

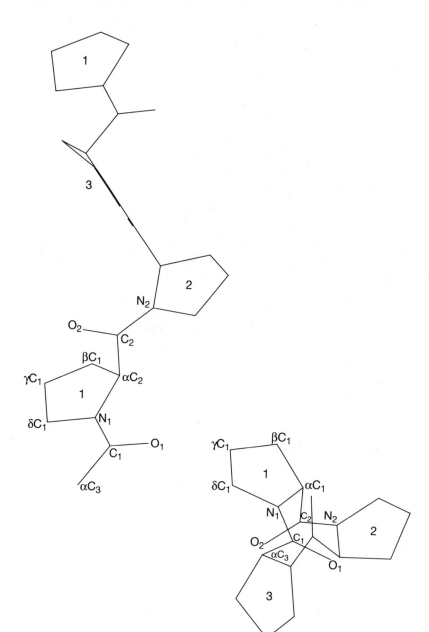

Figure A2.5. Helical structure of poly-L-proline. (Top) Seen normal to
the helix axis. (Bottom) Seen looking along that axis. Each residue is
related to its neighbor by a twist of 120° around and a shift of 3.1Å
along the helical axis. Three such helices are coiled around each
other in collagen. (Reproduced, by permission, from Cowan and
McGavin, 1955)

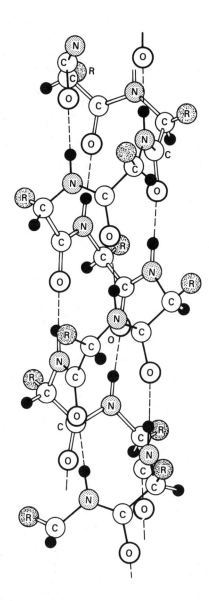

*Figure A2.6. Right-handed α-helix. It may be regarded as a spiral
staircase in which the amino acid residues form the steps. The height of
each step is 1.5Å, and the height of each turn is 5.4Å, making 3.6 steps
per turn. It takes 18 steps to return to a point exactly above the starting
point. The helix is held together by hydrogen bonds between the carbonyl
group of one residue and the imino group of the fourth residue down
along the chain. (Reproduced, by permission, from Creighton, 1983)*

other helices, designated as 2.2_7, 3.0_{10}, and 4.4_{16} had strain energies of 1 kcal per mole residue or less. However, Donohue considered only methyl groups as sidechains, and his conclusions might have been different for sidechains that are very large or interact strongly. Short 3.10_{10} and 4.4_{16} helical turns do occur in proteins, but 2.2_7 helices have not been found.

In a left-handed α-helix consisting of L-amino acids, the β-carbon atom is in *cis* position relative to the carbonyl group; in the right-handed one, it is in *cis* position relative to the imino group. The left-handed form gives rise to a short contact between any methyl or methylene group occupying the β position and the oxygen of the main chain carbonyl; it is therefore the less stable one and has not been encountered.

Pauling, Corey, and Branson discovered the α-helix by systematic model building. Its correctness was confirmed immediately from X-ray diffraction pictures of synthetic polypeptides. The α-helices could be recognized because the regular repeat of amino acid residues at axial intervals of 1.5Å gives rise to a strong X-ray reflection at 1.5Å spacing from planes perpendicular to the fiber axis. This reflection is of diagnostic value, because it is exhibited by the α-helix alone of all helical structures and is easily detected (Perutz, 1951).

Natural protein fibers belonging to the α-k.e.m.f. family all show the 1.5Å reflection characteristic of α-helices. (k stands for keratin of hair, horn, and nails; e for epidermin, the noncollagenous fiber of the skin; m for myosin of muscle; and f for fibrinogen, the blood-clotting protein.) The X-ray diagrams indicate that in these fibers two α-helices are coiled around each other (Crick, 1953). This happens because the sidechains of neighboring α-helices can interlock if the helices are inclined to each other at an angle of 20Å, but not if they are parallel, and such a mutual inclination is best achieved systematically in a coiled coil (see Chapter 3, page 105).

In globular proteins single polypeptide chains are folded into bundles of α-helices or pleated β-sheets or into combinations of them. Folding requires the chain to reverse its direction. Possible reverse turns fall into distinct classes, three of which are shown in Figure A2.7. Some of the classes impose restrictions on the kinds of residues allowed in certain positions. For example, formation of turn II is possible only if position 3 is occupied by glycine because a β-carbon atom at X would clash with the carbonyl oxygen atom of residue 2. This rule is of more than academic interest, because there exists at least one instance where substitution of another residue in that position of a protein causes cancer: the ATP-ase

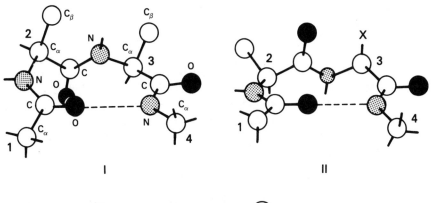

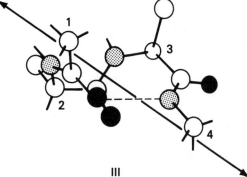

Figure A2.7. Three classes of reverse turns, each stabilized by a hydrogen bond between the carbonyl group of the first and the imino group of the fourth residue, but with different twist angles of the intervening amides. Class III is like a single turn of a 3_{10} helix; the arrow indicates the helix axis. (Reproduced, by permission, from Creighton, 1983)

coded for by the normal cellular *ras* proto-oncogene has a glycine in position 12 of a reverse turn. The ATP-ases coded for by certain of the malignant *ras* oncogenes differ from the normal ones only by the replacement of this single glycine by other residues. (See Chapter 8, page 205.)

Proteins are stabilized by a wide range of chemical interactions that can be characterized by the nature and energy of the forces between neighboring atoms and by their distances of separation. Atoms linked by covalent bonds share electrons; they are separated by between 1.0 and 1.7Å and their bond energies are of the order of tens of kcal/mol. In many kinds of chemical bonds separation of electronic charges gives rise to electric dipoles: in proteins the most important of these is the peptide bond in which the carbonyl oxygen carries a fractional negative charge and imide nitrogen a fractional positive charge. The mutual attraction of NH and OC groups gives rise to a chemical bond mediated by the hydrogen atom which is called a hydrogen bond. The N–H distance is fixed at 1.0Å; the H–O distance is variable between 1.5 and 2.0Å; the energy of hydrogen bonds varies between 1 and 5 kcal/mole. Strong hydrogen bonds are formed between ions of opposite charge, such as negatively charged carboxylate ions and positively charged ammonium, guanidinium, or imidazolium ions, especially if these ions are shielded from water. Formally neutral hydrocarbons attract each other by mutually induced dipoles. The separations between neighboring nonbonded atoms varies between 3 and 4Å. At a distance of 3.5Å, the bond energy between two aliphatic hydrocarbons is about 0.2 kcal/mole. Such bonds are said to be due to dispersion forces or van der Waals interactions, after the Dutch chemist who first showed that even noble gases do not obey the ideal gas law exactly, because colliding atoms momentarily stick together.

There is yet another force that stabilizes native protein structures. Hydrocarbons are sparingly soluble in water. In a classic paper, Frank and Evans had shown that the low solubility of hydrocarbons in water is due to an entropic effect, because water molecules tend to adhere to hydrocarbons, which reduces their mobility; therefore dissolving a hydrocarbon in water reduces the entropy of the system hydrocarbon plus water, which destabilizes it (Frank and Evans, 1945). Kauzmann argued that the same would happen when a protein unfolds and exposes its hydrocarbon sidechains to water, and that it is the water molecules' anarchic distaste for the orderly regimentation imposed upon them by the hydrocarbon sidechains that forces these sidechains to turn away from water and stick to each other in the interior of the protein (Kauzmann, 1959). This entropy-driven mechanism is known as the hydrophobic effect. Together with the van der Waals interactions, it provides about 0.024 kcal/mole of stabilization energy for every $Å^2$ of hydrophobic surface buried inside the protein.

Uncompensated electrical charges and strong dipoles tend to be immersed in the surrounding water because the dipolar water molecules screen and compensate their charges. Uncompensated interior charges make proteins unstable. For example, denaturation of proteins by alkali can be due to ionization of the sulphydryl groups of buried cysteines or of the phenolic hydroxyls of buried tyrosines. Denaturation by acid can be due to ionization of buried histidines (Perutz, 1974). (The pK_as of cysteines, tyrosines, and histidines in water are, respectively, 9.1–9.5, 9.5–10.0, and 6.3–6.8. The pK_a of a solute is the pH at which half the solute molecules are ionized.) One of the most harmful hemoglobin mutations consists in the replacement of an internal tyrosine by aspartate. The uncompensated negative charge of the carboxylate surrounds itself with water, which makes it a misfit and forces the globin to unfold (Lorkin et al., 1974).

Nucleic Acids
Nucleic acids are long-chain polymers, with a backbone in which an identical chemical pattern repeats at regular intervals, forming the links of the chain. This pattern consists of phosphates held between successive pentose residues. Each phosphate links the 3'-hydroxyl group of one pentose ring to the 5'-hydroxyl of the next. Attached to the 1' carbon atom of each pentose is one of four different kinds of nitrogenous bases: in DNA they are the two purines, adenine (A) and guanine (G), and the two pyrimidines, thymine (T) and cytosine (C) (Fig. A2.8). The four bases are the only variable constituents along the DNA chain; the specific sequence of the four bases constitutes the genetic information.

The total length of DNA and, by implication, the amount of genetic information carried by even the simplest organism is very great indeed. The smallest known chromosome is a single strand of DNA consisting of 5,500 nucleotides in the bacterial virus φX174. Most other DNAs isolated to date are double-stranded. The chromosome of T_2 bacteriophage contains $2 \cdot 10^5$ nucleotide pairs and that of a coli bacterium about 10^7. The number of nucleotide pairs in a single human germ cell is of the order of 10^9 and corresponds to an actual length of DNA of one meter!

In vivo, free DNA occurs only in certain viruses and bacteria. In higher organisms, it is associated with basic proteins: in the sperm heads of some species with protamine, which is rich in arginine, and in the nuclei of somatic cells with histones (see Chapter 3).

The structure of DNA consists of two chains that run in opposite directions and are coiled around each other to form a double helix (Fig. A2.9). It looks like a spiral staircase in which the bases form the steps and the phosphate ester chains provide the banisters. The bases with their large hydrophobic surfaces are neatly stacked together on the inside, while the negatively charged phosphate residues face the aqueous medium. Each step of the staircase

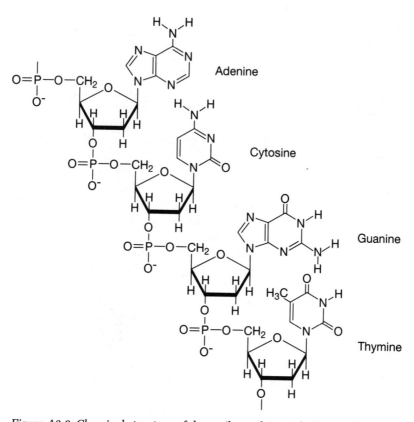

Figure A2.8. Chemical structure of deoxyribonucleic acid. The combination of one deoxyribose and one base is called a nucleoside; *that of one phosphate, one deoxyribose, and one base a (mono-)*nucleotide *or a* nucleosidemonophosphate. *Nucleoside di- or triphosphates have a chain of two or three phosphates attached to the deoxyribose ring. (Reproduced, by permission, from Perutz, 1962)*

consists of two bases, one from each of the chains, which are linked by hydrogen bonds, such that adenine is always linked to thymine and guanine to cytosine. This implies that the sequence of bases in the two chains making up the double helix is complementary, so that the sequence on one chain determines that on the other (Fig. A2.10). The rules of base pairing and the existence of complementary sequences in the two component chains arise from the geometry of the double helix and the stereochemical nature of the bases. X-ray analysis cannot provide information about the base sequence itself, because that sequence possesses no

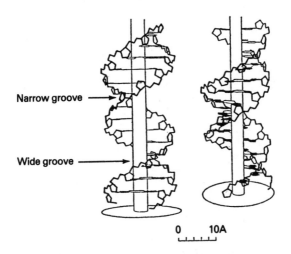

Figure A2.9. Diagram of Watson and Crick's double helix of DNA. The lines represent the phosphate ester chain. The disks between them indicate the paired bases. (Reproduced, by permission, from Perutz, 1962)

regular periodicity such as would be required to produce X-ray diffraction effects.

The bases lie with their planes approximately at right angles to the helical axis. There are 10 base pairs in each turn of the helix, separated by an axial distance of 3.4Å, and each turn has a height of 34Å, giving the helix an exact tenfold screw axis. The sense of the screw is right-handed. There are, however, additional symmetry elements that would make a model of DNA suitable for use as a staircase in a space ship. They have the effect of making it look the same when it is turned upside down. These elements are twofold rotation axes passing through each of the base pairs at right angles to the helical axis; each of them brings one of the chains into congruence with its partner of opposite polarity by a rotation of 180°. The axial spacing between the two chains alternates, as seen in Figure A2.9, leaving one wide and one narrow groove (Watson and Crick, 1953; Crick and Watson, 1954).

Figure A2.11 illustrates the chemical constitution of RNA. It is a polyester chain similar to DNA, from which it is distinguished by the presences of an additional hydroxyl group attached to the carbon (2′) of each pentose ring, and by the base uracil (U) which replaces thymine.

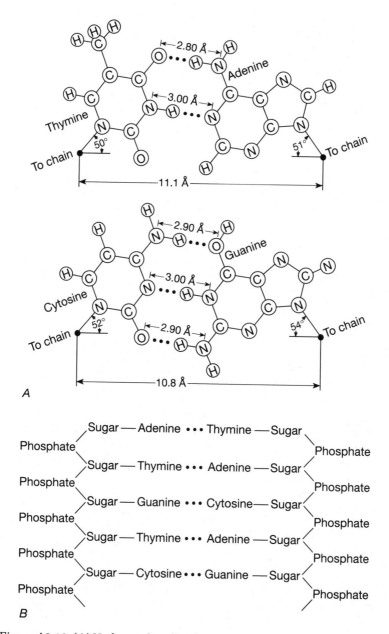

Figure A2.10. (A) Hydrogen bonding between the paired bases in DNA.
(B) Complementary sequences of bases in the two strands of the double
helix. Note that there is no preference for the direction of the pairs.
(Reproduced, by permission, from Perutz, 1962)

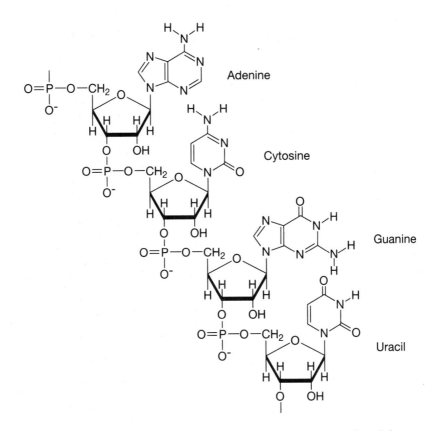

Figure A2.11. Chemical structure of ribonucleic acid. (Reproduced, by permission, from Perutz, 1962)

Further Reading

Creighton, T. E. 1983. *Proteins: Structures and Molecular Principles.* W. H. Freeman, New York.

Fersht, A. 1985. *Enzyme Structure and Mechanism.* 2nd ed. W. H. Freeman, New York.

Perutz, M. F. 1990. *Mechanisms of Cooperativity and Allosteric Control in Proteins.* Cambridge University Press, Cambridge, England.

Schulz, G. E., and R. H. Schirmer 1979. *Principles of Protein Structure.* Springer-Verlag, New York.

Stryer, L. 1988. *Biochemistry.* W. H. Freeman, New York.

Watson, J. D., N. H. Hopkins, J. W. Roberts, J. A. Steitz, and A. M. Weiner 1987. The importance of weaker chemical interactions. In *Molecular Biology of the Gene,* vol. 1. Benjamin/Cummings Publ., Menlo Park, CA, pp. 126–62.

References

Abate, C., L. Patel, F. J. Rauscher, III, and T. Curran 1990. Redox regulation of Fos and Jun DNA-binding activity in vitro. *Science* **247**:1157–61.

Abdel-Meguid, S. S., H. S. Shiek, W. W. Smith, H. E. Dayringer, B. N. Violand, and L. A. Bentle 1987. Three-dimensional structure of a genetically engineered variant of porcine growth hormone. *Proc. Nat. Acad. Sci. USA* **84**:6434–37.

Abraham, D. J., M. F. Perutz, and S. E. V. Phillips 1983. Physiological and X-ray studies of potential antisickling agents. *Proc. Nat. Acad. Sci. USA* **80**:324–28.

Acharya, R., E. Fry, D. Stuart, G. Fox, D. Rowland, and F. Brown 1989. The three-dimensional structure of the foot and mouth disease virus. *Nature* **337**:709–16.

Air, G. M., W. G. Laver, R. G. Webster, M. C. Els, and M. Luo 1989. Antibody recognition of the influenza virus neuraminidase. *Cold Spring Harbor Symp. Quant. Biol.* **54**:247–56.

Amit, A. G., R. A. Mariuzza, S. E. V. Phillips, and R. J. Poljak 1986. Three-dimensional structure of an antigen-antibody complex at 2.8Å resolution. *Science* **233**:747–53.

Andreeva, N. S., A. A. Fedorov, A. A. Gutshina, R. R. Riskulov, N. E. Schutzkeuer, and M. G. Safro 1978. *Molec. Biol. (USSR)* **12**:922–28.

Arai, K. -i., F. Lee, A. Miyjima, S. Miuatake, N. Arai, and T. Yokota 1990. Cytokines: Coordination of immune and inflammatory responses. *Ann. Rev. Biochem.* **59**:783–836.

Arents, G., R. W. Burlingame, B. -C. Wang, W. E. Love, and E. N. Moudrianakis 1991. The nucleosome core histone octamer at 3.1Å resolution: A tripartite protein assembly and a left-handed superhelix. *Proc. Nat. Acad. Sci. USA* **88**:10148–52.

Arnold, E., and M. G. Rossmann 1990. Analysis of the structure of a common cold virus, human rhinovirus 14, refined at a resolution of 3.0Å. *J. Mol. Biol.* **211**:763–801.

Arnone, A., R. E. Benesch, and R. Benesch 1977. Structure of human deoxyhemoglobin specifically modified with pyridoxal compounds. *J. Mol. Biol.* **115**:627–42.

Attwood, M. R., C. H. Hassall, A. Kröhn, G. Lawton, and S. Redshaw 1986. The design and synthesis of the angiotensin converting enzyme inhibitor Cilazapril and related bicyclic compounds. *J. Chem. Soc. Perkins Trans.* **I**:1011–19.

Badger, J., S. Krishnaswamy, M. J. Kremer, M. A. Oliveira, M. G. Rossmann, B. A. Heinz, R. R. Rueckert, F. J. Dutko, and M. A. McKinley 1989. Three-dimensional structures of drug-resistant mutants of human rhinovirus 14. *J. Mol. Biol.* **207**:163–74.

Baker, E. N., T. L. Blundell, J. F. Cutfield, S. M. Cutfield, E. J. Dodson, G. G. Dodson, D. M. Crowfoot Hodgkin, R. E. Hubbard, N. W. Isaacs, D. C. Reynolds, K. Sakabe, N. Sakabe, and N. M. Vijayan 1988. The structure of 2Zn insulin crystals at 1.5Å resolution. *Phil. Trans. Roy. Soc. B* **319**: 369–456.

Baldwin, E. T., I. T. Weber, R. St. Charles, J. -c. Xuan, E. Appella, M. Yamada, K. Matsushima, B. F. P. Edwards, G. M. Clore, A. M. Gronenborn, and A. Wlodawar 1991. Crystal structure of interleukin 8: Symbiosis of NMR and crystallography. *Proc. Nat. Acad. Sci. USA* **88**:502–06.

Baltimore, D. 1970. RNA-dependent DNA polymerase in visions of RNA tumor viruses. *Nature* **226**:1209–11.

Barbacid, M. 1987. *ras* Genes. *Ann. Rev. Biochem.* **56**:779–827.

Batchelor, J. R., and A. J. McMichael 1987. Progress in understanding HLA and disease associations. *Brit. Med. Bull.* **43**:156–83.

Baudin-Chich, V., J. Pagnier, M. Marden, N. Lacaze, J. Kister, O. Schaad, S. J. Edelstein, and C. Poyart 1990. Enhanced polymerization of recombinant human deoxyhemoglobin β6 Glu→Ile. *Proc. Nat. Acad. Sci. USA* **87**:1845–49.

Bawden, F. C., and N. W. Pirie 1938. Crystalline preparations of tomato bushy stunt mosaic virus. *Brit. J. Exp. Pathol.* **19**:251–60.

Beauclair, W. de 1949. *Verfahren und Geräte zur mehrdimensionalen Fouriersynthese.* Akademieverlag, Berlin.

Beese, L. S., and T. A. Steitz 1991. Structural basis for the 3′-5′ exonuclease activity of *Escherichia coli* DNA polymeraise I: A two metal ion mechanism. *EMBO J.* **10**:25–33.

Benesch, R., and R. E. Benesch 1981. Preparation and properties of hemoglobin modified with pyridoxal. *Methods in Enzymology* **76**:147–58.

Bentley, G. A., T. N. Bhat, G. Boulot, T. Fishman, J. Navaza, R. J. Poljak, M. -M. Riottot, and D. Tello 1989. Immunochemical and crystallographic studies of antibody D1.3 in its free, antigen-liganded, and idiotope-bound states. *Cold Spring Harbor Symp. Quant. Biol.* **54**:239–46.

Bentley, G. A., G. Boulet, M. M. Riottot, and R. J. Poljak 1990. Three-dimensional structure of an idiotope–anti-idiotope complex. *Nature* **348**:254–57.

Berg, J. M. 1988. Proposed structure for the zinc-binding domains from transcription factor IIIA and related proteins. *Proc. Nat. Acad. Sci. USA* **85**:99–102.

Bernal, J. D. 1939. The structure of proteins. *Nature* **143**:663–67.

Bernal, J. D., and D. Crowfoot 1934. X-ray photographs of crystalline pepsin. *Nature* **794**:133–34.

Bernal, J. D., I. Fankuchen, and D. P. Riley 1938. Structure of crystals of tomato bushy stunt virus. *Nature* **142**:1075.

Beutler, B., D. Greenwald, J. D. Hulmes, M. Chang, Y.-C. E. Pan, J. Mathison, R. Ulevitch, and A. Cerami 1985. Identity of tumour necrosis factor and the macrophage-secreted factor cachectin. *Nature* **316**:552–54.

Beutler, B. and A. Cerami 1988. Cachexia, shock and inflammation: a common mediator. *Ann. Rev. Biochem.* **57**:505–18.

Bhat, T. N., G. A. Butler, T. O. Fischman, G. Boulot, and R. J. Poljak 1990. Small rearrangements in structures of Fv and Fab fragments of antibody D1.3 on antigen binding. *Nature* **347**:483–84.

Billeter, M., A. D. Kline, W. Braun, R. Huber, and K. Wüthrich 1989. Comparison of high resolution structures of the α-amylase inhibitor tendamistat determined by nuclear magnetic resonance in solution and by X-ray diffraction in single crystals. *J. Mol. Biol.* **206**:677–87.

Billeter, M., Y. Q. Qian, G. Otting, M. Müller, W. J. Gehring, and K. Wüthrich 1990. Determination of the three-dimensional structure of the antennapedia homeodomain from *Drosophila* in solution by ^{1}H nuclear magnetic resonance spectroscopy. *J. Mol. Biol.* **214**:183–97.

Bjorkman, P. J., and P. Parham 1990. Structure and diversity of class I major histocompatibility complex molecules. *Ann. Rev. Biochem.* **59**:253–58.

Bjorkman, P. J., M. A. Saper, B. Samraoui, W. S. Bennett, J. L. Strominger, and D. C. Wiley 1987. The foreign antigen binding site and T-cell recognition regions of class I histocompatibility antigens. *Nature* **329**:512–17.

Blow, D. M., J. J. Birktoft, and B. S. Hartley 1969. Role of a buried acid group in the mechanism of action of chymotrypsin. *Nature* **221**:337–40.

Blow, D. M., and F. H. C. Crick 1959. The treatment of errors in the isomorphous replacement method. *Acta Cryst.* **12**:794–802.

Blow, D. M., J. Janin, and R. M. Sweet 1974. Mode of action of soybean trypsin inhibitor (Kunitz) as a model for specific protein–protein interaction. *Nature* **249**:54–57.

Blundell, T. L., and L. H. Johnson 1976. *Protein Crystallography.* Academic Press, New York.

Blundell, T. L., J. Cooper, S. I. Foundling, D. M. Jones, B. Atrash, and M. Szelke 1989. On the rational design of renin inhibition: X-ray studies of aspartic proteinases complexed with transition-state analogues. *Perspectives in Biochemistry,* ch. 13:83–89.

Blundell, T. L., S. M. Cutfield, J. F. Cutfield, E. J. Dodson, G. G. Dodson, D. C. Hodgkin, D. Mercola, and M. Vijayan 1971. Atomic positions in rhombohedral 2-zinc insulin crystals. *Nature* **231**:506–11.

Bode, W., I. Meyr, U. Baumann, R. Huber, S. R. Stone, and J. Hofsteenge 1989. The refined 1.9Å crystal structure of human α-thrombin: Interaction with D-Phr-Pro-Arg chloromethylketone and significance of the Tyr-Pro-Pro-Trp insertion segment. *EMBO J.* **8**:3467–75.

Boehm, T., K. Campbell, Y. Kanebo, M. F. Perutz, and T. H. Rabbitts 1991. The rhomobotin family of cysteine-rich LIM-domain oncogenes: Distinct members are involved in T-cell translocations to human chromosomes 11p15 and 11p13. *Proc. Nat. Acad. Sci. USA* **88**:4367–71.

Bokhoven, C., J. C. Schoone, and J. M. Bijvoet 1949. *Proc. Acad. Sci. Amst.* **52**:120–26.

Bos, J. L., E. R. Fearon, S. R. Hamilton, M. Verlaan-de Vries, J. H. van Boom, A. J. van der Erb, and B. Vogelstein 1987. Prevalence of colo-rectal *ras*-gene mutations in colo-rectal cancer. *Nature* **327**:293–95.

Bourne, H. R., D. A. Sanders, and F. McCormick 1991. The GTPase superfamily: Conserved structure and molecular mechanism. *Nature* **349**:117–27.

Bowie, J. U., R. Lüthy, and D. Eisenberg 1991. A method to identify protein sequences that fold into a known three-dimensional structure. *Science* **253**:164–70.

Boyle, W. J., T. Smeal, L. H. K. Defize, P. Angel, J. R. Woodgett, M. Karin, and T. Hunter 1991. Activation of protein kinase C decreases phosphorylation of c-Jun at sites that negatively regulate its DNA-binding activity. *Cell* **64**:573–84.

Bragg, Sir Lawrence 1975. The development of X-ray analysis. D. C. Phillips and H. Lipson, eds. G. Bell and Sons, London.

Bragg, W. H. 1915. X-rays and crystal structure. *Phil. Trans. Roy. Soc. (London) A* **215**:253–75.

Bragg, W. L. 1922. The diffraction of x-rays by crystals. Nobel Lectures: *Physics* 1901-21:370–82. Elsevier, Amsterdam 1967.

Brandhuber, B. J., T. Boone, W. C. Kenney, and D. B. McKay 1987. Three-dimensional structure of interleukin-2. *Science* **238**:1707–09.

Brange, J., U. Ribel, J. F. Hansen, G. Dodson, M. T. Hansen, S. Havelund, S. G. Melberg, F. Norris, K. Norris, L. Snel, A. R. Sorensen, and H. O. Voigt 1988. Monomeric insulins obtained by protein engineering and their medical implications. *Nature* **333**:679–82.

Brennan, R. G., S. L. Roderick, Y. Takeda, and B. W. Matthews 1990. Protein-DNA conformational changes in the crystal structure of λ-cro-operator complex. *Proc. Nat. Acad. Sci. USA* **87**:8165–69.

Brenner, S., L. Barnett, F. H. C. Crick, and L. Orgel 1961. The theory of mutagenesis. *J. Mol. Biol.* **3**:121–24.

Bricogne, G. 1976. Methods and programs for direct exploitation of geometric redundancies. *Acta Cryst.* **32**:832–47.

Broek, D., R. Bartlett, K. Crawford, and P. Nurse 1991. Involvement of p34^{cdc2} in establishing the dependency of S phase on mitosis. *Nature* **349**:388–93.

Brünger, A. T. 1992. Free R value: A novel statistical quantity for assessing the accuracy of crystal structures. *Nature* **355**:472–75.

Bunn, C. W. 1945. *Chemical Crystallography*. Clarendon Press, Oxford, England.

Burke, K. L., G. Dunn, M. Ferguson, D. P. Minor, and J. W. Almond 1988. Antigen chimeras of poliovirus as potential vaccines. *Nature* **332**:81–82.

Burley, S. K., and G. A. Petsko 1986. Amino-aromatic interactions in proteins. *FEBS Letts.* **203**:139–43.

Camerini, D., and B. Seed 1990. A CD4 domain important for HIV-mediated syncytium formation lies outside the virus binding site. *Cell* **60**:747–54.

Campbell, I. D., R. M. Cooke, M. Baron, T. S. Harvey and M. F. Tappin 1989. The solution structures of epidermal growth factor and transforming growth factor alpha. *Progess in Growth Factor Research* **1**:13–22

Camps, M. P., C. Murre, X.-h. Sun, and D. Baltimore 1990. A new homeobox gene contributes the DNA binding domain of the t(1,19) translocation protein in Pre-B ALL. *Cell* **60**:547–555. (Pre-B ALL stands for pre-B cell acute lymphocytic leukemia.)

Carlacci, L., K.-C. Chou, and G. M. Maggiora 1991. A heuristic approach to predicting the tertiary structure of bovine somatotrophin. *Biochemistry* **30**:4389–98.

Caspar, D. L. D., and A. Klug 1962. Physical principles in the construction of regular viruses. *Cold Spring Harbor Symp. Quant. Biol.* **27**:1–24.

"Central Dogma Reversed" 1970. *Nature* **226**:1198–99.

Cerundolo, V., T. Elliot, J. Elvin, J. Bastin, H.-G. Rammensee, and A. Townsend 1991. The binding affinities and dissociation rates of peptides for Class I MHC molecules. *Eur. J. Biochem.* **21**:2069–75.

Chatterjee, R., E. V. Welty, R. Y. Walder, S. L. Pruitt, P. H. Rogers, A. Arnone, and J. A. Walder 1986. Isolation and characterisation of a new hemoglobin derivative cross-linked between the α-chains (lysine $99\alpha_1$–lysine $99\alpha_2$). *J. Biol. Chem.* **261**:9929–37.

Choi, H.-K., L. Tong, W. Minor, P. Dumas, U. Boege, M. G. Rossmann, and G. Wengler 1991. Structure of sindbis virus core protein reveals a chymotrypsin-like serine proteinase and the organisation of the virion. *Nature* **354**:37–42.

Chothia, C., M. Levitt, and D. Richardson 1977. Structure of proteins: Packing of α-helices and pleated sheets. *Proc. Nat. Acad. Sci. USA* **74**:4130–34.

Christianson, C., J. Tranum-Jensen, J. Carlson, and J. Vinten 1991. A model of the quaternary structure of human placental insulin receptor deduced from electron microscopy. *Proc. Nat. Acad. Sci. USA* **88**:249–52.

Christianson, D. W., P. R. David, and W. N. Lipscomb 1987. Mechanism of carboxypeptidase A: Hydration of a ketonic substrate analogue. *Proc. Nat. Acad. Sci. USA* **84**:1512–15.

Churchill, M., and M. Suzuki 1989. "SPKK" motifs prefer to bind to DNA at A/T-rich sites. *EMBO J.* **8**:4889–95.

Clore, G. M., and A. M. Gronenborn 1991. Comparison of the solution nuclear magnetic resonance and crystal structures of interleukin 8. Possible implications for the mechanism of receptor binding. *J. Mol. Biol.* **217**:611–20.

Clore, G. M., E. Apella, M. Yamada, K. Matsushima, and A. M. Gronenborn 1990. Three-dimensional structure of interleukin 8 in solution. *Biochemistry* **29**:1689–96.

Clore, G. M., P. T. Wingfield, and A. M. Gronenborn 1991. High-resolution three-dimensional structure of interleukin 1β in solution by three- and four-dimensional nuclear magnetic resonance spectroscopy. *Biochemistry* **30**:2315–23.

Coll, M., J. Aymani, G. A. van der Marel, J. J. van Boom, A. Rich, and A. H.-J. Wang 1989. Molecular structure of the netropsin-d (CGCGATATCGCG) complex: DNA conformation in an alternating AT segment. *Biochemistry* **28**:310–20.

Colman, P. M., J. N. Varghese, and W. G. Laver 1983. Structure of the catalytic and antigenic sites in influenza virus neuraminidase. *Nature* **303**:41–44.

Cooper, J. Personal communication.

Cowan, P. M., and S. McGavin 1955. Structure of poly-L-proline. *Nature* **176**:501–03.

Craig, E. A., and C. A. Gross 1991. Is hsp 70 the cellular thermometer? *Trends in Biochem. Sci.* **16**:135–40.

Creighton, T. E. 1983. *Proteins: Structures and molecular principles.* W. H. Freeman, New York.

Crick, F. H. C. 1952. Is α-keratin a coiled coil? *Nature* **170**:882.

Crick, F. H. C. 1953. The Fourier transform of a coiled coil. *Acta Cryst.* **6**:685–89.

Crick, F. H. C., and A. Rich 1955a. Structure of polyglycine II. *Nature* **176**:780–81.

Crick, F. H. C., and A. Rich 1955b. The structure of collagen. *Nature* **176**:915–16.

Crick, F. H. C., and J. D. Watson 1954. The complementary structure of deoxyribonucleic acid. *Proc. Roy. Soc. (London)* A.**223**:80–96.

Crick, F. H. C., L. Barnett, S. Brenner, and R. J. Watts-Tobin 1961. General nature of the genetic code for proteins. *Nature* **192**:1227–32.

Crowther, R. A. 1971. Procedures for three-dimensional reconstruction of spherical viruses by Fourier synthesis from electron micrographs. *Phil. Trans. Roy. Soc. Lond.* B.**261**:221–30.

Cullis, A. F., H. Muirhead, M. F. Perutz, M. G. Rossmann, and A. C. T. North 1961. The structure of haemoglobin VIII. A three-dimensional Fourier synthesis at 5.5Å resolution: Determination of the phase angles. *Proc. Roy. Soc. (London)* A.**265**:15–38.

Cunningham, B. A., J. J. Hemperly, B. A. Murray, E. A. Prediger, C. Brackenbury, and G. M. Edelman 1987. Neural cell adhesion molecule: Structure, immunoglobulin-like domains, cell surface modulation, and alternative RNA splicing. *Science* **236**:799–805.

Cunningham, B. C., M. Ultsch, A. M. de Vos, M. G. Mulherrin, K. R. Clauser, and J. A. Wells 1991. Dimerization of the extracellular domain of the human growth hormone receptor by a single hormone molecule. *Science* **254**:821–25.

Curtis, B. M., S. R. Presnell, S. Srinivasan, H. Sassenfeld, R. Klinke, E. Jeffery, D. Cosman, C. J. March, and F. E. Cohen 1991. Experimental and theoretical studies of the three-dimensional structure of human interleukin 4. *Proteins* **11**:111–19.

Cushman, D. W., H. S. Cheung, E. F. Sabo, and M. A. Ondetti 1977. Design of potent competitive inhibitors of angiotensin-converting enzyme. Carboxyalkanoyl and mercaptalkanoyl amino acids. *Biochemistry* **16**:5484–91.

Davies, D. R., and E. A. Padlan 1990. Antibody-antigen complexes. *Ann. Rev. Biochem.* **59**:439–73.

Davies, J. F., Z. Hostomska, Z. Hostomsky, S. Jordan, and D. A. Matthews 1991. Crystal structure of the ribonuclease H domain of HIV-1 reverse transcriptase. *Science* **252**:88–95

Davis, M. M., and P. J. Bjorkman 1988. T-cell antigen receptor genes and T-cell recognition. *Nature* **334**:395–402.

DeCaprio, J. A., J. W. Ludlow, J. Figge, J.-Y. Shew, C.-M. Huang, S.-H. Lee, E. Marsilio, E. Paucha, and D. M. Livingstone 1988. SV40 large tumour antigen forms a specific complex with the product of the retinoblastoma susceptibility gene. *Cell* **54**:275–83.

Deisenhofer, J., O. Epp, K. Miki, R. Huber, and H. Michel 1985. Structure of the protein subunits in the photosynthetic reaction centre of *Rhodopseudomonas viridis* at 3Å resolution. *Nature* **318**:618–24.

Derewenda, U., Z. Derewenda, E. J. Dodson, G. G. Dodson, and Xiao Bing 1991. X-ray analysis of the single chain B29-A1 peptide-linked insulin molecule. *J. Mol. Biol.* **220**:425–33.

Desjarlais, R. L., G. L. Seibel, I. D. Kuntz, P. S. Furth, J. C. Alvarez, P. R. Ortiz de Montellano, D. L. DeCamp, L. M. Babe, and C. S. Craik 1990. Structure-based design of nonpeptide inhibitors specific for the human immunodeficiency virus 1 protease. *Proc. Nat. Acad. Sci. USA* **87**:6644–48.

Dickerson, R. E., J. C. Kendrew, and B. E. Strandberg 1961. The crystal structure of myoglobin: Phase determination to a resolution of 2Å by the method of isomorphous replacement. *Acta Cryst.* **14**:1188–95.

Djuran, M. I., E. L. M. Lempers, and J. Redijk 1991. Reactivity of chloro- and aqua(diethylenetriamine)platinum II ions with glutathione, S-methyl-glutathione, and guanosine-5'-monophosphate in relation to antitumour activity and toxicity of platinum complexes. *Inorg. Chem.* **30**:2648–52.

Donohue, J. 1953. Hydrogen-bonded helical configurations of the polypeptide chain. *Proc. Nat. Acad. Sci. USA* **39**:470–78.

Driscoll, P. C., J. G. Cyster, I. D. Campbell, and A. F. Williams 1991. Structure of domain 1 of rat T lymphocyte CD2 antigen. *Nature* **353**:762–65.

Duncan, A. R., and G. Winter 1988. The binding site for C1q on IgG. *Nature* **332**:738–40.

Dunitz, J. D. 1979. *X-ray Analysis and the Structure of Organic Molecules.* Cornell University Press, Ithaca, N.Y.

Dyckerhoff, H., and G. Tewes 1933. Über die Adsorption von Pepsin an Eiweiss. *Hoppe Seyl. Z. Physiol. Chem.* **215**:93–120.

Ealick, S. E., W. J. Cook, S. Vijak-Kumar, M. Carson, T. L. Nagabushan, P. P. Trotta, and C. E. Bugg 1991. Three-dimensional structure of recombinant human interferon-γ. *Science* **252**:698–702.

Eck, M. J., and S. R. Sprang 1989. The structure of tumor necrosis factor-α at 2.6Å resolution: Implications for receptor binding. *J. Biol. Chem.* **264**:17595–605.

Eigenbrot, C., M. Randal, C. Quan, J. Burnier, L. O'Connell, E. Rinderknecht, and A. A. Kossiakoff (1991). X-ray structure of human relaxin at 1.5Å. Comparison to insulin and implications for receptor binding determinants. *J. Mol. Biol.* **221**:15–21.

Ellis, R. W., D. DeFeo, T. Y. Shih, M. A. Gonda, H. A. Young, N. Tsuchida, D. R. Lowy, and E. M. Scolnick 1981. The p21 *src* genes of Harvey and Kirsten sarcoma viruses originate from divergent members of a family of normal vertebrate genes. *Nature* **292**:506–11.

Engel, J. and D. J. Prockop 1991. The zipper-like folding of collagen triple helices and the effects of mutations that disrupt the zipper. *Ann. Rev. Biophys. Chem.* **20**:137–52.

Farrell, P. J., G. C. Sen, M. F. Dubois, L. Ratner, E. Slattery, and P. Lengyell 1978. Interferon action: Two distinct pathways for inhibitors of protein synthesis by double-stranded RNA. *Proc. Nat. Acad. Sci. USA* **75**:5893–97.

Fermi, G., and M. F. Perutz 1981. *Atlas of Molecular Structures in Biology,* Vol. 2. *Haemoglobin and Myoglobin.* Clarendon Press, Oxford, England.

Fermi, G., M. F. Perutz, B. Shaanan, and R. Fourme 1984. The crystal structure of human deoxyhaemoglobin at 1.74Å resolution. *J. Mol. Biol.* **175**:159–74.

Fersht, A. 1985. *Enzyme Structure and Mechanism.* 2nd ed. W. H. Freeman, New York, p. 339.

Finch, J. T., A. Klug, and R. Leberman 1970. The structures of turnip crinkle and tomato bushy stunt viruses II. The surface structure: Dimer clustering patterns. *J. Mol. Biol.* **50**:215–22.

Finzel, B. C., L. L. Clancy, D. R. Holland, S. W. Muchmore, K. D. Watenpaugh, and H. M. Einspahr 1989. Crystal structure of recombinant human interleukin-1β at 2.0Å resolution. *J. Mol. Biol.* **209**:779–91.

Fitzgerald, P. M. D., B. M. McKever, J. F. Van Middlesworth, J. P. Springer, J. C. Heimbach, C.-T. Leu, W. K. Herber, R. A. F. Dixon, and P. L. Darke 1990. Crystallographic analysis of a complex between human immunodeficiency virus type 1 protease and acetylpepstatin at 2.0Å resolution. *J. Biol. Chem.* **265**:14209–19.

Flaherty, K. M., C. De Luca-Flaherty, and D. B. McKay 1990. Three-dimensional structure of the ATP-ase fragment of a 70 kD heat shock protein. *Nature* **346**:623–28.

Fleming, A. 1929. On the antibacterial action of cultures of a *Penicillium*, with special references to their use in the isolation of *B. influenzae*. *British J. Exp. Path.* **10**:226–36.

Flynn, G., P. G. Chappell, and J. E. Rothman 1989. Peptide binding and release by proteins implicated as catalysts of protein assembly. *Science* **245**:385.

Flynn, G. C., J. Pohl, M. T. Flocco, and J. E. Rothman 1991. Peptide binding specificity of molecular chaperone BiP. *Nature* **353**:726–30.

Frank, H. S., and M. W. Evans 1945. Free volume and entropy in condensed systems III: Entropy in binary liquid mixtures; partial modal entropy in dilute solutions; structure and dynamics in aqueous electrolytes. *J. Chem. Phys.* **13**:507–32.

Freyd, G., S. K. Kim, and H. R. Horwitz 1990. Novel cysteine-rich motif and homeodomain in the product of the Caenorhabditis elegans cell lineage gene lin-11. *Nature* **344**:876–79.

Gale, E. F., E. Cundliffe, P. E. Reynolds, M. H. Richmond, and K. J. Waring 1981. *The Molecular Basis of Antibiotic Action*. 2nd ed. Wiley, London.

Garret, T. P. J., M. A. Saper, P. J. Bjorkman, J. L. Strominger, and D. C. Wiley 1989. Specificity pockets for the side chain of peptide antigens in HLA-AW68. *Nature* **342**:692–96.

Gehring, W. J. 1985. The molecular basis of development. *Sci. Am.,* October:**137–46.**

Gehring, W. J. 1987. Homeoboxes in the study of development. *Science* **236**:1245–52.

Gianelli, F., P. M. Green, K. A. High, S. Sommer, D. P. Lillicrap, M. Ludwig, K. Oleg, P. H. Reitsma, M. Gossens, A. Yoshioka, and G. G. Brownlee 1991. Haemophilia B: Database of point mutations and short additions and deletions, 2nd edition. *Nucl. Acids Res.* **19**:2193–219.

Gilbert, W., and B. Muller-Hill 1966. Isolation of the *Lac* repressor. *Proc. Nat. Acad. Sci. USA* **56**:1891–98.

Giranda, V. L., M. S. Chapman, and M. G. Rossmann 1990. Modelling of the human intercellular adhesion molecule-1, the human rhinovirus major group receptor. *Proteins* **7**:227–33.

Goff, S. P. 1990. Retroviral reverse transcriptase: Synthesis, structure and function. *J. Acquired Immune Deficiency Syndromes* **3**:817–31.

Goodford, P. J. 1985. A computational procedure for determining energetically favourable binding sites on biologically important macromolecules. *J. Med. Chem.* **28**:849–57.

Green, D. W., V. M. Ingram, and M. F. Perutz 1954. The structure of haemoglobin IV. Sign determination by the isomorphous replacement method. *Proc. Roy. Soc.* **A225**:287–307.

Guinier, A. 1963. *X-ray Diffraction in Crystals, Imperfect Crystals and Amorphous Bodies*. W. H. Freeman, San Francisco.

Hale, G., M. R. Clark, R. Marcus, G. Winter, M. J. S. Dyer, J. M. Phillips, L. Riechmann, and H. Waldmann 1988. Remission induction in non-Hodgkin lymphoma with reshaped human monoclonal antibody Campath-1H. *Lancet*:1394–99.

Härd, T., E. Kellenbach, R. Boelens, B. A. Maler, K. Dahlman, L. P. Freedman, J. Carlstedt-Duke, K. R. Yamamoto, J.-Ä. Gustafsson, and R. Kapstein 1990. Solution structure of the glucocorticoid receptor DNA-binding domain. *Science* **249**:157–60.

Harrison, S. C. 1991. What do viruses look like? *Harvey Lectures* **87**:127–52.

Harrison, S. C. 1983. *Adv. Virus Res.* **28**:175–240.

Harrison, S. C., A. J. Olson, C. E. Schutt, F. K. Winckler, and G. Bricogne 1978. Tomato bushy stunt virus at 2.9Å resolution. *Nature* **276**:368–73.

Harvey, J. J. 1964. An unidentified virus which causes the rapid production of tumors in mice. *Nature* **204**:1104–05.

Harvey, T. S., A. J. Wilkinson, M. J. Tappin, R. M. Cooke and I. D. Campbell 1991. The solution structure of human transforming growth factor. *Eur. J. Biochem.* **198**:555–62.

Hassall, C. H., A. Kröhn, C. H. Moody, and W. A. Thomas 1982. The design of a new group of angiotensin-converting enzyme inhibitors. *FEBS Letts.* **147**:175–79.

Hassall, C. H., A. Kröhn, C. H. Moody, and W. A. Thomas 1984. The design and synthesis of new triazolo- and pyridazo-pyridazin derivatives as inhibitors of angiotensin converting enzyme. *J. Chem. Soc. Perkins Trans.* **1**:155–64.

Hassall, C. H., W. H. Johnson, A. J. Kennedy, and N. A. Roberts 1985. A new class of inhibitors of human leukocyte elastase. *FEBS Letts.* **183**:201–05.

Henderson, R., J. M. Baldwin, T. A. Ceska, F. Zemlin, E. Beckmann, and K. H. Downing 1990. Model for the structure of *bacteriorhodopsin* based on high resolution electron cryo-microscopy. *J. Mol. Biol.* **213**:899–929.

Herron, J. N., X.-M. He, A. L. Gibson, E. W. Voss, and A. B. Edmundson 1987. Crystal structure of a murine F_{ab} fragment with specificity for single-stranded DNA. *Fed. Proc.* **43**:1626.

Hill, C. S., J. M. Rimmer, B. N. Green, J. T. Finch, and J. O. Thomas 1991. Histone–DNA interactions and their modulation by phosphorylation of -Ser-Pro-X-Lys/Arg- motifs. *EMBO J.* **10**:1939–48.

Hogle, J. M., M. Chow, and D. J. Filman 1985. Three-dimensional structure of poliovirus at 2.9Å resolution. *Science* **229**:1358–65.

Holmes, K. C., and D. M. Blow 1965. The use of X-ray diffraction in the study of protein and nucleic acid structure. *Methods of Biochemical Analysis* **13**:133–239.

Holmgren, A., and C.-I. Brändén 1989. Crystal structure of chaperone protein PapD reveals an immunoglobulin fold. *Nature* **342**:248–51.

Hsu, I. N., L. T. Delbaere, M. N. G. James, and T. Hofmann 1977. Penicillopepsin from *Penicillium janthinellum*. Crystal structure at 2.8Å and sequence homology with porcine pepsin. *Nature* **266**:140–45.

Hua, Q. X., S. E. Shoelson, M. Kochoyan, and M. A. Weiss 1991. Receptor binding redefined by a structural switch in a mutant human insulin. *Nature* **354**:238–41.

Huisman, T. H. J., A. K. Brown, G. D. Efremor, J. B. Wilson, C. A. Reynolds, R. Uy, and L. L. Smith 1971. Hemoglobin Savannah (B6(24)β Glycine →Valine): An unstable variant causing anemia with inclusion bodies. *J. Clin. Invest.* **50**:650–59.

Isaacs, A., and J. Lindemann 1957. Virus interference I. The interferon. *Proc. Roy. Soc. B.* **147**:258–67.

Jacob, F. 1982. *The Possible and the Actual.* University of Washington Press, Seattle, WA.

Jacob, F. 1988. *The Statue Within: An Autobiography.* Basic Books, New York.

Jacob, F., and J. Monod 1961. Genetic regulatory mechanisms in the synthesis of proteins. *J. Mol. Biol.* **3**:318–56.

Jain, S. C., and H. M. Sobel 1972. Stereochemistry of actinomycin binding to DNA. I. Refinement and further structural details of the actinomycin-deoxyguanosine complex. *J. Mol. Biol.* **68**:1–20.

Janin, J. and C. Chothia 1990. The structure of protein–protein recognition sites. *J. Biol. Chem.* **265**:16027–30.

Johnston 1987. A fungal gene regulator mechanism: The GAL genes of saccharomyces cerevisiae. *Microbiol. Rev.* **51**:486–96.

Jones, E. Y., D. I. Stuart, and N. P. C. Walker 1989. Structure of tumour necrosis virus. *Nature* **338**:225–28.

Jordan, S. R., and C. O. Pabo 1988. Structure of the lambda complex at 2.5Å resolution: Details of the repressor–operator interactions. *Science* **242**:895–99.

Kabsch, W., H. G. Mannherr, D. Suck, E. F. Pai, and K. C. Holmes 1990. Atomic structure of the actin: DNA-ase I complex. *Nature* **347**:37–43.

Karle, I. L., J. L. Flippen-Anderson, S. Agarwalla, and P. Balaram 1991. Crystal structure of [Leu[1]]zervamicin, a membrane ion-channel peptide: Implications for gating mechanisms. *Proc. Nat. Acad. Sci. USA* **88**:5307–5311

Karlsson, O., S. Thor, T. Norberg, H. Ohlson, and T. Edlund 1990. Insulin gene enhancer binding protein Lsl-1 is a member of a novel class of proteins containing both a homeo- and a Cys-His domain. *Nature* **344**:879–82.

Katayanagi, K., M. Miyagawa, M. Matsushima, M. Ishikawa, S. Kanaya, M. Ikehara, T. Matsusaki, and K. Morikawa 1990. Three-dimensional structure of ribonuclease H from *E. coli. Nature* **347**:306–09.

Kauzmann, W. 1959. Some factors in the interpretation of protein denaturation. *Adv. Protein Chem.* **14**:1–64.

Kay, L. E., M. G. Clore, A. Bax, and A. M. Gronenborn 1990. Four-dimensional heteronuclear triple-resonance NMR spectroscopy of interleukine-1β in solution. *Science* **249**:411–14.

Kennedy, M. A., R. Gonzales-Sarmiento, U. R. Kees, E. Lampert, N. Dear, T. Boehm, and T. H. Rabbitts 1991. HOX11, a homeobox-containing T-cell oncogene on human chromosome 10q24. *Proc. Nat. Acad. Sci.* **88**:8900–04.

Kim, S., T. J. Smith, M. S. Chapman, M. G. Rossmann, D. C. Pevear, F. J. Dutko, P. J. Felock, G. D. Diana, and M. A. McKinley 1989. Crystal structure of human rhinovirus serotype 1 (HRV1A). *J. Mol. Biol.* **210**:91–111.

Kirsten, W. H., and L. A. Mayer 1976. Morphological responses to a murine erythroblastosis virus. *J. Nat. Cancer. Inst.* **39**:311–19.

Kissinger, C. R., B. Liv, E. Martin-Blanco, T. B. Kornberg, and C. O. Pabo 1990. Crystal structure of an *engrailed* homeodomain-DNA complex at 2.8Å resolution: A framework for understanding homeodomain–DNA interactions. *Cell* **63**:579–90.

Klug, A., and D. Rhodes 1987. Zinc-fingers: A novel protein motif for nucleic acid recognition. *Trends Biochem. Sci.* **12**:464–69.

Kopka, M. L., C. Yoon, D. Goodsell, C. Pjura, and R. E. Dickerson 1985. The molecular origin of DNA-drug specificity in netropsin and distamycin. *Proc. Nat. Acad. Sci. USA* **82**:1376–80.

Koshland, D. E. 1959. In *The Enzymes,* 2nd ed. vol. 1, P. D. Boyer, H. Lardy, and K. Myrbäck, eds. Academic Press, New York, pp. 305–46.

Kornberg, R. D., and A. Klug 1981. The nucleosome. *Sci. Am.* **214**:48–60.

Kouzarides, T., and E. Ziff 1988. The role of the leucine zipper in the fos–june interaction. *Nature* **336**:646–51.

Kühlbrandt, W., and D. N. Wang 1991. Three-dimensional structure of plant light-harvesting complex determined by electron crystallography. *Nature* **350**:130–34.

Labeit, S., D. P. Barlow, M. Gautel, T. Gibson, J. Holt, C.-L. Hsieh, U. Francke, K. Leonard, J. Wardale, A. Whiting, and J. Trinick 1990. A regular pattern of two types of 100-residue motif in the sequence of titin. *Nature* **345**:276.

La Cour, T. F. M., J. Nyborg, S. Thirup, and B. F. C. Clark 1985. Structural details of the binding of guanosine phosphate to elongation factor Tu from *E. coli* by X-ray crystallography. *EMBO J.* **4**:2385–88.

Lalezari, I., P. Lalezari, C. Poyart, M. Marden, J. Kister, B. Bohn, G. Fermi, and M. F. Perutz 1990. New effectors of human hemoglobin: Structure and function. *Biochemistry* **29**:1515–23.

Lalezari, I., S. Rahbar, P. Lalezari, G. Fermi, and M. F. Perutz 1988. LR16, a compound with potent effects on the oxygen affinity of hemoglobin, on blood cholesterol and on low density lipoprotein. *Proc. Nat. Acad. Sci. USA* **85**:6117–21.

Landschulz, W. H., P. F. Johnson, and S. L. McKnight 1988. The leucine zipper: A hypothetical structure common to a new class of proteins. *Science* **240**:1759–64.

Langridge, R., W. E. Seeds, H. R. Wilson, C. W. Hooper, M. H. F. Wilkins, and D. M. Hamilton 1957. Molecular structure of deoxyribonucleic acid (DNA). *J. Biophys. Biochem. Cytol.* **3**:767–78.

Langs, D. A., G. D. Smith, C. Courseille, G. Précigoux, and M. Hospital 1991. Monoclinic uncomplexed double-stranded, antiparallel, left-handed $\beta^{5.6}$-helix $(\uparrow\downarrow\beta^{5.6})$ structure of gramicidin A: Alternate patterns of helical association and deformation. *Proc. Nat. Acad. Sci. USA* **88**:5345–49.

Lapetto, R., T. Blundell, T. Hemmings, J. Overington, A. Wilderspin, S. Wood, J. R. Merson, P. J. Whittle, D. E. Danleigh, K. F. Geoghegan, S. J. Hawrylik, S. E. Lee, K. J. Scheld, and P. M. Hobart 1989. X-ray analysis of HIV-1 proteinase at 2.7Å resolution confirms structural homology among retroviral enzymes. *Nature* **342**:299–342.

Lavelle, D., J. Ducksworth, E. Ewes, G. Gomes, M. Keller, P. Heller, and J. De Simone 1991. A homeodomain protein binds to γ-globin regulatory sequences. *Proc. Nat. Acad. Sci. USA* **88**:7318–22.

Lee, M. S., G. P. Gippert, K. V. Soman, D. A. Case, and P. E. Wright 1989. Three-dimensional solution structure of a single zinc finger DNA-binding domain. *Science* **245**:635–37.

Leibrock, J., F. Lottspeich, A. Hohn, M. Hofer, B. Hengerer, P. Masiakowski, H. Thoenen, and Y.-A. Barde 1989. Molecular cloning and expression of brain-derived neurotrophic factor. *Nature* **341**:149–52.

Lerman, L. S. 1961. Structural considerations in the interaction of DNA and acridines. *J. Mol. Biol.* **3**:18–30.

Lesk, A. M., and C. Chothia 1984. Mechanisms of domain closure in proteins. *J. Mol. Biol.* **174**:175–91.

Levi-Montalcini, R. 1988. In *Praise of Perfection.* Basic Books, Inc., New York, N. Y., p. 148.

Levitt, M., and M. F. Perutz 1988. Aromatic rings act as hydrogen bond acceptors. *J. Mol. Biol.* **201**:751–54.

Liddington, R. C., Y. Yan, J. Moulai, R. Sahli, T. L. Benjamin, and S. C. Harrison 1991. Structure of simian virus 40 at 3.8Å resolution. *Nature* **354**:278–84.

Lindskog, S., L. E. Henderson, K. K. Kannan, A. Liljas, P. O. Nyman, and B. Strandberg 1971. Carbonic anhydrase, in Boyer, P. D. (ed.), *The Enzymes,* vol. 5., pp. 587-665, Academic Press, New York and London.

Lipscomb, W. N. 1970. Structure and mechanism in the enzymatic activity of carboxypeptidase A and relations to chemical sequence. *Acc. Chem. Res.* **3**:81–89.

Loebermann, H., R. Tokuoka, J. Deisenhofer, and R. Huber 1984. Human α_1-antitrypsin inhibitor. Crystal structure analysis of two crystal modifications, molecular analysis of two crystal modifications, molecular model and preliminary analysis of the implications for function. *J. Mol. Biol.* **177**:531–56.

Lomas, D., D. Evans, and R. Carrell 1992. Molecular mechanism of antitrypsin accumulation in the liver. *Lancet* (in press).

Looker, D., D. Abbott-Brown, P. Cozart, S. Durfee, S. Hoffmann, A. J. Matthews, J. Miller-Roehrich, S. Shoemaker, S. Trimble, G. Fermi, N. H. Komiyama, K. Nagai, and G. L. Stetler 1992. A novel human recombinant haemoglobin designed for use as a blood substitute. *Nature* (In press.)

Lorkin, P. A., H. Pietschmann, H. Brannsteiner, and H. Lehmann 1974. Structure of haemolobin Wien β130 (H8) tyrosine-aspartic acid: An unstable haemoglobin variant. *Acta Haematol.* **51**:351–61.

Luisi, B. F., X. W. Xu, Z. Otwinowski, L. P. Freedman, K. R. Yamamoto, and B. P. Sigler 1991. Crystallographic analysis of the interaction of the glucocorticoid receptor with DNA. *Nature* **352**:497–505.

Lüthy, R., J. U. Bowie, and D. Eisenberg 1992. Assessment of protein models wit 3D profiles. *Nature* **356**:83–85.

Lwoff, A. 1966. The prophage and I. In *The Phage and Molecular Biology*, J. Cairns, G. S. Stent, and J. D. Watson, eds. Cold Spring Harbor Laboratory, Long Island.

Madden, D. R., J. C., Gorga, J. L. Strominger, and D. C. Wiley 1991. The structure of HLA-B27 reveals nonamer self-peptides bound in an extended conformation. *Nature* **353**:321–25.

Mader, S., V. Kumar, H. de Verneuil, and P. Chambon 1989. Three amino acids of the oestrogen receptor are essential to its ability to distinguish an oestrogen from a glucocorticoid-responsive element. *Nature* **338**:271–74.

Markland, W., and A. E. Smith 1987. Mutants of the polyomavirus middle-T antigen. *Biochem. Biophys. Acta* **907**:299–321.

Marrack, P., and J. Kapler 1986. The T-cell and its receptor. *Sci. Am.*, February:28.

Martin, A., C. Wychowski, T. Coudere, R. Ce, J. Hogle, and M. Girard 1988. Engineering a poliovirus 2 antigenic site on a type 1 capsid results in a chimeric virus which is neurovirulent to mice. *EMBO. J.* **7**:2839–47.

Martin, A. J. P., and R. L. M. Synge 1945. *Adv. in Prot. Chem.* **2**:1.

Matthews, B. W. 1988. Structural basis of the action of thermolysin and related zinc peptidases. *Acc. Chem. Res.* **21**:333–39.

Matthews, B. W., B. P. Sigler, B. Henderson, and D. M. Blow 1967. Three-dimensional structure of toxyl-α-chymotrypsin. *Nature* **214**:652–56.

Matthews, D. A., R. A. Alden, J. T. Bolin, S. T. Freer, R. Hamlin, W. G. J. Hol, R. L. Kisliuk, E. J. Pastore, L. T. Plante, N. Xuong, and J. Kraut 1978. Dihydrofolate reductase from *Lactobacillus casei*. "X-ray structure of the enzyme methotrexate complex." *J. Biol. Chem.* **253**:6946–54.

McDonald, N. Q., R. Lapatto, J. Murray-Rust, J. Gunning, A. Wlodawer, and T. L. Blundell 1991. New protein fold revealed by a 2.3Å resolution crystal structure of nerve growth factor. *Nature* **354**:411–14.

McGuire, E. A., R. D. Hockett, K. M. Pollack, M. F. Bartholdi, S. O. O'Brien, and S. J. Korsmeyer 1989. *Mol. Cell. Biol.* **9**:2124–32.

Milburn, M. V., L. Tong, A. M. DeVos, A. Brünger, S. Yamaizumi, S. Nishimura, and S.-H. Kim 1990. Molecular switch for signal transduction: Structural differences between active and inactive forms of proto-oncogenic *ras* proteins. *Science* **247**:939–45.

Miller, J., A. D. McLachlan, and A. Klug 1985. Repetitive zinc-binding domains in the protein transcription factor IIIA from *Xenopus* oocytes. *EMBO J.* **4**:1609–14.

Miller, M., B. K. Sathyanarayana, M. V. Toth, G. R. Marshal, L. Clawson, L. Selk, J. Schneider, S. B. H. Kent, and A. Wlodawer 1989. Complex of synthetic HIV-1 protease with a substrate-based inhibitor at 2.3Å resolution. *Science* **246**:1149–52.

Miller, R. A., D. G. Maloney, R. Warnke, and R. Levy 1982. Treatment of B cell lymphoma with monoclonal anti-idiotype antibodies. *N. Engl. J. Med.* **306**:517–19.

Minor, P. D., M. Ferguson, K. Katrak, D. Wood, H. John, J. Howlett, G. Dunn, K. Burke, and J. W. Almond 1990. Antigenic structure of chimeras of type 1 and type 2 poliovirus involving antigenic site 1. *J. Gen. Virology* **71**:2543–51.

Monzinga, A. F., and B. W. Matthews 1984. Binding of N-carboxymethyl dipeptide inhibitors to thermolysin determined by X-ray crystallography: A novel class of transition-state analogues for zinc peptidase. *Biochemistry* **23**:5724–29.

Moore, M. H., W. N. Hunter, B. L. d'Estaintot, and O. Kennard 1989. DNA–drug interactions. The crystal structure of d(CGTACG) complexed with daunomycin. *J. Mol. Biol.* **206**:693–705.

Moreno, S., and P. Nurse 1990. Substrates for $p34^{cdc2}$: In vivo veritas? *Cell* **61:549**–51.

Morimoto, H., H. Lehmann, and M. F. Perutz 1971. Molecular pathology of human haemoglobin: Stereochemical interpretation of abnormal oxygen affinities. *Nature* **232**:408–13.

Mottonen, J., A. Strand, J. Symersky, R. M. Sweet, D. E. Danley, K. F. Geoghegan, R. D. Gerard, and E. J. Goldsmith 1992. Structural basis of latency in plasminogen activator. *Nature*. (In press.)

Murray, M. G., R. J. Kuhn, M. Arita, N. Kawamura, A. Nomoto, and E. Wimmer 1988. Poliovirus type 1/type 3 antigenic hybrid virus constructred *in vitro* elicits type 1 and type 3 neutralizing antibodies in rabbits and monkeys. *Proc. Nat. Acad. Sci. USA* **85**:3203–07.

Nagai, K., C. Oubridge, T. H. Jessen, J. Li, and P. R. Evans 1990. Crystal structure of the RNA-binding domain of the U1 small nuclear ribonucleoprotein A. *Nature* **348**:515–20.

Navia, M. A., P. M. D. Fitzgerald, B. M. McKeever, C.-T. Leu, J. C. Heimbach, W. K. Herber, I. S. Sigal, P. L. Darke, and J. P. Springer 1989. Three-dimensional structure of aspartyl protease from human immuno-deficiency virus HIV1. *Nature* **337**:615–20.

Neidle, S., A. Achari, G. L. Taylor, H. M. Berman, H. L. Carrel, J. P. Glusker, and W. C. Stalling 1977. Structure of dinucleoside phosphate-drug complex as a model for nucleic acid–drug interaction. *Nature* **269**:304–07.

Nicol, C. S., G. L. Mayer, and S. M. Russel 1986. Structural features of prolactin and growth hormones that can be related to their biological properties. *Endocrine Reviews* **7**:169–85

Nourse, J., J. D. Mellentin, N. Galili, J. Wilkinson, E. Stanbridge, S. D. Smith, and M. L. Cleary 1990. Chromosomal translocation t(1,19) results in synthesis of a homeobox fusion mRNA that codes for a potential chimeric transcription factor. *Cell* **60**:535–45.

Nunes, A. C., and B. P. Schoenborn 1973. Dichloromethane and myoglobin function. *Mol. Pharmacol.* **9**:835–39.

O'Shea, E. K., J. D. Klem, P. S. Kim, and T. Alber 1991. X-ray structure of the GCN4 leucine zipper, a two-stranded parallel coiled coil. *Science* **254**:539–44.

O'Shea, E. K., R. Rutkowski, W. F. Stafford, III, and P. S. Kim 1989. Preferential heterodimer formation by isolated leucine zippers from *Fos* and *Jun*. *Science* **245**:646–48.

Oas, T. G., L. P. McIntosh, E. K. O'Shea, W. M. Dahlquist, and P. S. Kim 1990. Secondary structure of a leucine zipper determined by nuclear magnetic resonance spectroscopy. *Biochemistry* **29**:2891–94.

Ofir, R., V. J. Dwarki, D. Rashid, and I. M. Verma 1990. Phosphorylation of the C-terminus of *Fos* protein is required for transcriptional transrepression of the *c-fos* promoter. *Nature* **348**:80–82.

Ostermann, J., A. L. Horwick, W. Neuport, and F. Ulrich-Hartl 1989. Protein folding in mitochondria requires complex formation with hsp60 and ATP hydrolysis. *Nature* **341**:125–30.

Otting, G., Y. Q. Qian, M. Billeter, M. Müller, M. Affolter, W. J. Gehring, and K. Wüthrich 1990. Protein-DNA contacts in the structure of a homeodomain-DNA complex determined by nuclear magnetic resonance spectroscopy in solution. *EMBO J.* **9**:3085–92.

Owen, M. C., S. O. Brennan, J. H. Lewis, and R. W. Carrell 1983. Mutation of antitrypsin to antithrombin α_1-antitrypsin Pittsburgh (358 Met→ Arg): a fatal bleeding disorder. *New Eng. J. Med.* **309**:694–98.

Page, G. S., A. G. Mosser, J. M. Hogle, D. J. Filman, R. R. Rueckert, and M. Chow 1988. Three-dimensional structure of poliovirus serotype 1 neutralizing determinants. *J. Virol.* **62**:1781–94.

Pai, E. F., W. Kabsch, U. Krengel, K. C. Holmes, J. John, and A. Wittinghofer 1989. Structure of guanine-nucleotide-binding domain of the Ha-*ras* oncogene product p21 in the triphosphate conformation. *Nature* **341**:209–14.

Párraga, G., S. J. Horvath, A. Eisen, W. E. Taylor, L. Hood, E. T. Young, and R. E. Klevit 1988. Zinc-dependent structure of a single-finger domain of yeast ADR1. *Science* **241**:1489–92.

Patchett, A. A., E. Harris, E. W. Tristram, M. J. Wyvratt, M. T. Wu, D. Taub, E. R. Peterson, T. J. Ikeler, J. ten Broeke, L. G. Payne, D. L. Ondeyka, E. D. Thorsett, W. J. Greenlee, N. S. Lohr, R. D. Hoffsomer, H. Joshua, W. V. Ruyle, J. W. Rothrock, S. D. Aster, A. L. Maycock, F. M. Robinson, and R. Hirschmann 1980. A new class of angiotensin-converting enzyme inhibitors. *Nature* **288**:280–83.

Patterson, A. L. 1934. A Fourier series method for the determination of the components of interatomic distances in crystals. *Phys. Rev.* **46**:372–76.

Paucker, K., K. Cantell, and W. Henle 1962. Quantitative studies on suspended L cells III. Effect of interfering viruses and interferon on the growth of cells. *Virology* **17**:324–34.

Pauling, L. 1948. Nature of forces between large molecules of biological interest. *Nature* **161**:707–09.

Pauling, L., and R. B. Corey 1951. The pleated sheet, a new layer configuration of the polypeptide chain. *Proc. Nat. Acad. Sci. USA* **37**:251–56.

Pauling, L., R. B. Corey, and H. R. Branson 1951. The structure of proteins: Two hydrogen-bonded helical configurations of the polypeptide chain. *Proc. Nat. Acad. Sci. USA* **37**:205–11.

Pavletich, N. P., and C. O. Pabo 1991. Zinc finger-DNA recognition: Crystal structure of a ZIF268-DNA complex at 2.1Å resolution. *Science* **252**:809–16.

Pelham, H. R. B. 1986. Speculation on the functions of the major heat shock proteins and glucose-regulated proteins. *Cell* **46**:959–61.

Perutz, M. F. 1951. New X-ray evidence on the configuration of the polypeptide chain. *Nature* **167**:1053–58.

Perutz, M. F. 1974. Mechanism of denaturation of haemoglobin by alkali. *Nature* **247**:341–44.

Perutz, M. F. 1962. *Proteins and Nucleic Acids.* Elsevier, Amsterdam.

Perutz, M. F. 1989. Mechanisms of cooperativity and allosteric regulation in proteins. *Quart. Rev. Biophys.* **22**:139–236.

Perutz, M. F. 1990. Mechanisms regulating the reactions of human haemoglobin with oxygen and carbon monoxide. *Ann. Rev. Physiol.* **52**:1–25.

Perutz, M. F., and C. Poyart 1983. Bezafibrate lowers the oxygen affinity of haemoglobin. *Lancet* **II**:881–82.

Perutz, M. F., G. Fermi, D. J. Abraham, C. Poyart, and E. Bursaux 1986. Hemoglobin as a receptor of drugs and peptides: X-ray studies of the stereochemistry of binding. *J. Amer. Chem. Soc.* **108**:1065–78.

Petska, S., J. A. Langer, K. C. Zoon, and C. E. Samuel 1987. Interferons and their actions. *Ann. Rev. Biochem.* **56**:727–77.

Pirie, N. W. 1937. The meaninglessness of the terms life and living. In *Perspectives in Biochemistry,* J. Needham and D. E. Green, eds. Cambridge University Press, pp. 11–22. Cambridge, England.

Poljak, R. J., L. M. Amzel, H. P. Avery, B. L. Chen, R. P. Phizakerly, and F. Saul 1973. Three-dimensional structure of the F_{ab}' fragment of a human immunoglobulin at 2.8Å resolution. *Proc. Nat. Acad. Sci. USA* **70**:3305–10.

Priestle, J. P., H.-P. Schär, and M. G. Grütter 1989. Crystallographic refinement of interleukin 1β at 2.0Å resolution. *Proc. Nat. Acad. Sci. USA* **86**:9667–71.

Ptashne, M. 1967. Isolation of the λ-phage repressor. *Proc. Nat. Acad. Sci. USA* **57**:306–13.

Ptashne, M. 1986. *A Genetic Switch: Gene Control and Phage λ.* Cell Press and Blackwell Scientific Publications. Oxford, England.

Qian, Y. Q., M. Billeter, G. Otting, M. Muller, W. J. Gehring, and K. Wuthrich 1989. The structure of the antennapedia homeodomain determined by NMR spectroscopy in solution: Comparison with prokaryotic repressors. *Cell* **59**:573–80.

Quiqley, G. J., G. Ughetto, G. Van der Marel, J. H. Van Boom, A. H.-J. Wang, and A. Rich 1986. Non-Watson-Crick GC and AT base pairs in a DNA-antibiotic complex. *Science* **232**:1255–58.

Ransone, L. J., and I. M. Verma 1990. Nuclear proto-oncogenes *Fos* and *Jun.* *Ann. Rev. Cell Biol.* **6**:539–57.

Rasmussen, R., D. Benvegnu, E. K. O'Shea, P. S. Kim, and T. Alber 1991. X-ray scattering indicates that the leucine zipper is a coiled coil. *Proc. Nat. Acad. Sci. USA* **88**:561–64.

Rayment, I., T. S. Baker, D. L. D. Caspar, and W. T. Murakami 1982. Polyoma virus capsid structure at 22.5Å resolution. *Nature* **295**:110–15.

Reddy, E. P., R. K. Reynolds, E. Santos, and M. Barbacid 1982. A point mutation is responsible for the acquisition of transforming properties by T24 human bladder carcinoma. *Nature* **300**:149–52.

Redfield, C., L. J. Smith, J. Boyd, G. M. P. Lawrence, R. G. Edwards, R. A. G. Smith, and C. M. Dobson 1991. Secondary structure and topology of human interleukin-4 in solution. *Biochemistry* **30**: 11029–35.

Reedjik, J., A. M. N. Fietinger-Schepman, J. T. van Oosterum, and P. van de Putte 1987. Platinum amine coordination compounds as anti-tumour drugs: Molecular aspects of the mechanism of action. *Structure and Bonding* **67**:53–89.

Rich, A. and F. H. C. Crick 1955. The structure of collagen. *Nature* **176**: 915–16.

Richmond, J. J., J. T. Finch, B. Rushton, D. Rhodes, and A. Klug 1984. Structure of the nucleosome core particle at 7Å resolution. *Nature* **311**:532–37.

Riechmann, L., M. Clark, H. Waldmann, and G. Winter 1988. Reshaping human antibodies for therapy. *Nature* **332**:323–27.

Rippmann, F., W. R. Taylor, J. B. Rothbard, and M. N. Green 1991. A hypothetical model for the peptide binding domain of hsp70 based on the peptide binding domain of HLA. *Embo. J.* **10**:1053–59.

Roberts, M. M., J. L. White, M. G. Grutter, and R. M. Burnett 1986. Three-dimensional structure of the adenovirus major coat protein hexon. *Science* **232**:1149–51.

Robertson, J. M. 1936. An X-ray study of the phthalocyanins. Part II: Quantitative structure determination of the metal-free compound. *J. Chem. Soc.*:1195–1209.

Robertson, J. M., and I. Woodward 1940. An X-ray study of the phthalocyanins. Part IV: Direct quantitative analysis of the platinum compound. *J. Chem. Soc.*:36–48.

Robey, E., and R. Axel 1990. CD4: Collaborator in immune recognition and HIV infection. Minireview. *Cell* **60**:697–700.

Rodríguez-Tébar, A., G. Dechant, and Y.-A. Barde 1991. Neurotrophins: Structural relatedness and receptor interactions. *Phil. Trans. Roy. Soc. (London) B* **331**:255–58.

Rossmann, M. G. 1960. The accurate determination of the position and shape of heavy-atom replacement groups in proteins. *Acta Cryst.* **13**:221–26.

Rossmann, M. G. 1989. The structure of antiviral agents that inhibit uncoating when complexed with viral capsids. *Antiviral Research* **11**:3–14.

Rossmann, M. G., and J. E. Johnson 1989. Icosahedral RNA virus structure. *Ann. Rev. Biochem.* **58**:533–73.

Rossmann, M. G., and A. C. Palmenberg 1988. Conservation of the putative receptor attachment site in picornaviruses. *Virology* **164**:373–82.

Rould, M. A., J. J. Perona, and T. A. Steitz 1991. Structural basis of anticodon loop recognition by glutaminyl synthease. *Nature* **352:**213–18.

Rydel, T. J., K. C. Ravichandran, A. Tulinsky, W. Bode, R. Huber, C. Roitsch, and J. W. Fenton 1990. The structure of a complex of recombinant hirudin and human α-thrombin. *Science* **249:**277–80.

Ryle, A. P., F. Sanger, L. F. Smith, and R. Kitai 1955. *Biochem. J.* **60:**541–50.

Ryu, S.-E., P. D. Kwong, A. Trunch, T. G. Porter, G. Arthos, M. Rosenberg, X. Dai, N.-h. Xuong, R. Axel, R. W. Sweet, and W. A. Hendrickson 1990. Crystal structure of an HIV-binding recombinant fragment of human CD4. *Nature* **348:**419–25.

Salunke, D. M., D. L. D. Caspar, and R. L. Garcea 1986. Self-assembly of purified polyoma capsid protein VP1. *Cell* **46:**895–904.

Samraovi, B., P. J. Bjorkman, and D. C. Wiley 1988. A hypothetical model of the foreign antigen binding site of class II histocompatibility antigens. *Nature* **332:**845–50.

Saper, M. A., P. J. Bjorkman, and D. C. Wiley 1991. Refined structure of the human histocompatibility antigen HLA-2 at 2.6Å resolution. *J. Mol. Biol.* **219:**277–319.

Sasaki, K., Y. Yamano, S. Bardhan, N. Iwai, J. J. Murray, M. Hasegawa, Y. Matsuda, and T. Inagami 1991. Cloning and expression of a complementary DNA encoding a bovine adrenal angiotensin II type-1 receptor. *Nature* **351:**230–33.

Sauer, R. T., R. R. Yokum, R. F. Doolittle, M. Lewis, and C. O. Pabo 1982. Homology among DNA-binding proteins suggests use of a conserved super-secondary structure. *Nature* **298:**447–51.

Sauter, N. K., M. D. Bednarski, B. A. Wurzburg, J. E. Hanson, G. M. Whitesides, J. J. Skehel, and D. C. Wiley 1989. Hemaglutinins from two influenza virus variants bind to sialic acid derivatives with millimolar dissociation constants: A 500-MHz proton nuclear magnetic resonance study. *Biochemistry* **28:**8388–96.

Schechter, A. N., C. T. Noguchi, and G. P. Rodgers 1987. Sickle cell disease. In *The Molecular Basis of Blood Diseases,* G. Stammatoyannopoulos, A. W. Nienhuis, P. Leder, and P. W. Majerus, eds. W. B. Saunders Co., Philadelphia.

Schiffer, M., R. L. Girling, K. R. Ely, and A. B. Edmundson 1973. Structure of a λ-type Bence-Jones protein at 3.5Å resolution. *Biochemistry* **12:**4620–31.

Schoenborn, B. P. 1976. Dichloromethane as an antisickling agent in sickle cell hemoglobin. *Proc. Nat. Acad. Sci. USA* **73:**4195–99.

Schultz, S. C., G. C. Shields, and T. A. Steitz 1991. Crystal structure of a CAP-DNA complex: The DNA is bent by 90°. *Science* **253:**1001–07.

Schwabe, J. W. R., D. Neuhaus, and D. Rhodes 1990. Solution structure of the DNA binding domain of the oestrogen receptor. *Nature* **348:**458–61.

Senda, T., S. Matsuda, H. Kurihara, K. T. Nakamura, G. Kawano, H. Shimizu, H. Mizuno, and Y. Mitsui 1990. Three-dimensional structure of recombinant murine interferon-β. *Proc. Japan Acad. B.* **66:**77–80. Also *Protein Engineering 5 1992.* (In press.)

Sheldrick, G. M. 1990. Phase annealing in SHELX-90: Direct methods for larger structures. *Acta Cryst.* **A46:**467–73.

Sherman, S. E., D. Gibson, A. H.-J. Wang, and S. J. Lippard 1985. X-ray structure of the major adduct of the anticancer drug cisplatin with DNA: cis-[Pt(NH₃)₂{d(pGpG)}], *Science* **230:**412–17.

Sherman, S. E., D. Gibson, A. H.-J. Wang, and S. J. Lippard 1988. Crystal and molecular structure of cis-[Pt(NH₃)₂(d(pGpG)]: The principal adduct formed by cis-diammine dichloroplatinum (II) with DNA. *J. Amer. Chem. Soc.* **110:**7368–81.

Shieh, H.-S., H. M. Berman, M. Dabrov, and S. Neidle 1980. The structure of a drug-deoxynucleoside complex: Generalized conformational behaviour of intercalation complexes with RNA and DNA. *Nucleic Acid Res.* **8:**85–97.

Shih, T. Y., M. O. Weeks, H. A. Young, and E. M. Scolnick 1979a. Identification of a sarcoma virus-coded phosphoprotein in nonproducer cells transformed by Kirsten or Harvey murine sarcoma virus. *Virology* **96:**64–79.

Shih, T. Y., M. O. Weeks, H. A. Young, and E. M. Scolnick 1979b. p21 of Kirsten sarcoma virus is thermolabile in a viral mutant temperature sensitve for the maintenance of transformation. *J. Virol.* **31:**546–56.

Sielecki, A. R., A. A. Fedorov, A. Boodhoo, N. S. Andreeva, and M. N. G. James 1990a. Molecular and crystal structures of monoclinic porcine pepsin refined at 1.8Å resolution. *J. Mol. Biol.* **214:**143–70.

Sielecki, A. R., K. Hayakawa, M. Fujinaga, M. E. P. Murphy, M. Fraser, A. K. Muir, C. T. Carilly, J. A. Lewicki, J. D. Baxter, and M. N. G. James 1990b. Structure of recombinant human renin, a target for cardiovascular-active drugs, at 2.5Å resolution. *Science* **243:**1346–51.

Skehel, J. J., D. J. Stevens, R. S. Daniels, A. R. Douglas, M. Knossow, I. A. Wilson, and D. C. Wiley 1984. A carbohydrate sidechain on hemaglutinins of Hong Kong influenza virus inhibits recognition by monoclonal antibody. *Proc. Nat. Acad. Sci. USA* **81:**1779–83.

Skehel, J. J., B. C. Barnet, D. S. Burt, R. S. Daniels, A. R. Douglas, C. M. Graham, J. Hodgson, M. Knossow, K. H. G. Mills, P. F. Riska, D. B. Thomas, W. Weis, D. C. Wiley, and N. G. Wrigley 1989. Immune recognition of influenza virus haemoglutinin. *Phil. Trans. Roy. Soc. B* **323:**479–85.

Smith, K. M. 1935. A new disease of the tomato. *Ann. Appl. Biol.* **22:**731–41.

Smith, T. J., M. J. Kremer, M. Luo, G. Vriend, E. Arnold, G. Kamer, M. G. Rossmann, M. A. McKinley, G. D. Diana, and M. J. Otto 1986. The site of attachment in human rhinovirus 14 for antiviral agents that inhibit uncoating. *Science* **233:**1286–93.

Sobell, H. M., C.-C. Tsai, S. C. Jain, and S. G. Gilbert 1977. Visualization of drug–nucleic acid interactions at atomic resolution. III unifying structural concepts in understanding drug–DNA interactions and their broader implications in understanding protein–DNA interactions. *J. Mol. Biol.* **114:**333–65.

Sprang, S. R., and M. J. Eck 1992. The 3-D structure of TNF. In *The Tumour Necrosis Factors: The Molecules and Their Emerging Role in Medicine*, B. Beutler. ed. Raven Press, New York.

St. Charles, R., D. A. Walz, and B. F. P. Edwards 1989. The three-dimensional structure of bovine platelet factor 4 at 3.0Å resolution. *J. Biol. Chem.* **264**:2092–99.

Staunton, D. E., V. J. Merluzzi, R. Rothlein, R. Barton, S. D. Marlin, and T. A. Springer 1989. A cell surface adhesion molecule, ICAM1, is the major surface receptor for rhinoviruses. *Cell* **56**:849–53.

Stein, P. E., A. G. W. Leslie, J. T. Finch, and R. W. Carrell 1991. Crystal structure of uncleaved ovalbumin at 1.95Å resolution. *J. Mol. Biol.* **221**:941–59.

Stein, P. E., A. G. W. Leslie, J. T. Finch, W. G. Turnell, P. J. McLoughlin, and R. W. Carrell 1990. Crystal structure of ovalbumin as a model for the reactive centre of serpins. *Nature* **347**:99–102.

Steitz, T. A. 1991. Aminoacyl-tRNA synthetases: Structural aspects of evolution and tRNA recognition. *Current Opinion in Structural Biology* **1**:139–43.

Stewart, P. L., R. M. Burnett, M. Cyrclaff, and S. D. Fuller 1991. Image reconstruction reveals the complex molecular organisation of adenovirus. *Cell* **67**:145–54.

Stryer, L. 1988. *Biochemistry,* 3rd ed. W. H. Freeman, New York, p. 787.

Subramanian, E., I. D. A. Swan, M. Liu, D. R. Davies, J. A. Jenkins, I. J. Tickle, and T. L. Blundell 1977. Homology among acid proteases: Comparison of crystal structures at 3Å resolution of acid proteases from *Rhizopus chinensis* and *Endothia parasitica*. *Proc. Nat. Acad. Sci. USA* **74**:556–59.

Suguna, K. R. R. Bolt, E. A. Padlan, E. Subramanian, S. Sheriff, G. H. Cohen, and D. R. Davies 1987. Structure and refinement at 1.8Å resolution of the aspartic protease from *Rhizopus chinensis*. *J. Mol. Biol.* **196**:877–90.

Sundquist, W. I., S. J. Lippard, and D. B. Stollar 1987. Monoclonal antibodies to DNA modified with *cis* or transdiamminedichloroplatinum (II). *Proc. Nat. Acad. Sci. USA* **84**:8225–29.

Suzuki, M. 1989. SPXX, a frequent sequence motif in gene regulatory proteins. *J. Mol. Biol.* **207**:61–84.

Suzuki, M. 1991. The DNA-binding motif, SPKK, and its variants. In *Nucleic Acids and Molecular Biology,* vol. 5., F. Eckstein, D. M. J. Lilley, eds. Springer-Verlag, Berlin, pp. 126–140.

Suzuki, M., H. Sohma, M. Yazawa, K. Yagi, and S. Ebashi 1990. Histone H1 kinase specific to the SPKK motif. *J. Biochem.* **108**:356–64.

Tabin, C. J., S. M. Bradley, C. I. Bargman, R. A. Weinberg, A. G. Papageorge, E. M. Scolnick, R. Dhar, D. R. Lowy, and E. H. Chang 1982. Mechanism of activation of a human oncogene. *Nature* **300**:143–49.

Taparowsky, E., Y. Suard, O. Fasano, K. Shimizu, M. Goldfarb, and M. Wigler 1982. Activation of the T24 bladder carcinoma transforming gene is linked to a single amino acid change. *Nature* **300**:762–64.

Tanaka, I., K. Appelt, J. Dijk, S. W. White, and K. S. Wilson 1984. 3Å resolution structure of a protein with histone-like properties in prokaryotes. *Nature* **310:**376–81.

Temin, H. M. 1963. The effects of actinomycin D on growth of Rous sarcoma virus *in vitro*. *Virology* **20:**577–82.

Temin, H. M., and S. Mizutani 1970. RNA-dependent DNA polymerase in virions of Rous sarcoma virus. *Nature* **226:**1211–13.

Thoenen, H. 1991. The changing scene of neurotrophic factors. *Trends in Neurosci.* **14:**165–70.

Townes, P. L., and J. Holtfreter 1955. Directed movement and selective adhesion of embryonic amphibian cells. *J. Exp. Zool.* **128:**53–120.

Travers, P. 1990. One hand clapping. *Nature* **348:**393–94.

Treisman, J., P. Gouczy, M. Vashista, E. Harris, and C. Desplan 1989. A single amino acid can determine DNA binding specificity of homeodomain proteins. *Cell* **59:**553–62.

Tsai, C.-C., S. C. Jain, and H. M. Sobell 1975. X-ray crystallographic visualization of drug–nucleic acid intercalative binding: Structure of an ethidium-dinucleoside monophosphate crystalline complex: ethidium: 5-iodouridylyl(3'-5') adenosine. *Proc. Nat. Acad. Sci. USA* **72:**628–32.

Tsai, C.-C., S. C. Jain, and H. M. Sobell 1977. Visualization of drug-nucleic acid interactions at atomic resolution: I. Structure of an ethidium/ dinucleoside monophosphate crystalline complex, ethidium: 5-iodouridylyl(3'-5') adenosine. *J. Mol. Biol.* **114:**301–15.

Tulip, W. R., J. N. Varghese, A. T. Baker, A. van Donkelaar, W. G. Laver, R. G. Webster, and P. M. Colman 1991. Refined atomic structures of N9 subtype influenza virus neuraminidase and escape mutants. *J. Mol. Biol.* **221:**487–497.

Tulip, W. R., J. N. Varghese, R. G. Webster, G. M. Air, W. G. Laver, and P. M. Colman 1989. Crystal structure of neuraminidase-antibody complexes. *Cold Spring Harbor Symp. Quant. Biol.* **54:**257–64.

Ulrich, A., J. R. Bell, E. Y. Chen, R. Herrera, L. M. Petruzzelli, T. J. Dull, A. Gray, L. Coussens, Y.-C. Liao, M. Tsubokawa, A. Mason, P. H. Seeburg, C. Grunfeld, O. M. Rosen, and J. Ramachandran 1985. Human insulin receptor and its relationship to the tyrosine kinase family of oncogenes. *Nature* **313:**756–61.

Umesono, K., and R. M. Evans 1989. Determinants of target gene specificity for steroid/thyroid hormone receptors. *Cell* **57:**1139–46.

Valegard, K., L. Liljas, K. Fridborg, and T. Unge 1990. The three-dimensional structure of the bacterial virus MS2. *Nature* **345:**36–41.

Valentine, R. C., and H. G. Peirera 1965. Antigens and structure of the adenovirus. *J. Mol. Biol.* **13:**13–20.

Van der Veer, J. L., and J. Reedjik 1988. Investigating antitumour drug mechanisms. *Chem. in Britain* **24:**775–80.

Van Oostrum, J., and R. M. Burnett 1985. Molecular composition of the adenovirus type 2 virion. *J. Virol.* **56:**439–48.

Varghese, J. N. and P. M. Colman 1991. Three-dimensional structure of the neuraminidase of influenza virus A/Tokyo/3/67 at 2.2Å resolution. *J. Mol. Biol.* **221**:473–486.

Varghese, J. N., W. G. Laver, and P. M. Colman 1983. Structure of the influenza virus glycoprotein antigen neuraminidase at 2.9Å resolution. *Nature* **303**:35–40.

Vasák, M., E. Wörgötter, G. Wagner, J. H. R. Kagi, and K. Wüthrich 1987. Metal coordination in rat liver metallocyanin-2 prepared with and without reconstitution of the metal clusters, and comparison with rabbit liver metallocyanin-2. *J. Mol. Biol.* **196**:711–19.

Vilcek, J., and T. H. Lee 1991. Tumor necrosis factor: New insights into the molecular mechanisms of action. *J. Biol. Chem.* **266**:7313–16.

Vinson, C. R., P. B. Sigler, and S. L. McKnight 1989. Scissor-grip model for DNA recognition by a family of leucine zipper proteins. *Science* **246**:911–16.

de Vos, A. M., M. Ultzch, and A. A. Kossiakoff 1992. Human growth hormone and extracellular domain of its receptor: Crystal structure of the complex. *Science* **255**:306–12.

Vulliamy, T., P. Mason, L. Luzzatto 1992. The molecular basis of glucose 6-phosphate dehydrogenase deficiency. *Trends in Genetics.* (In press.)

Walder, J. A., R. Y. Walder, and A. Arnone 1980. Development of antisickling compounds that chemically modify hemoglobin S specifically within the 2,3-diphosphoglycerate binding site. *J. Mol. Biol.* **141**:195–216.

Waldmann, T. A. 1989. The multi-subunit interleukin-2 receptor. *Ann. Rev. Biochem.* **58**:875–914.

Waldmann, T. A. 1991. The interleukin-2 receptor. *J. Biol. Chem.* **266**:2681–84.

Waldschmidt-Leitz, E. 1933. The chemical nature of enzymes. *Science* **78**:189–90.

Wallace, B. A., and K. Ravikumar 1988. The gramicidin pore: Crystal structure of a cesium complex. *Science* **241**:182–87.

Wang, A. H.-J., G. Ughetto, G. J. Quigley, and A. Rich 1987. The interaction between an anthracyclin antibiotic and DNA: Molecular structure of daunomycin complexed to d(CGTACG) at 1.2Å resolution. *Biochemistry* **26**:1152–63.

Wang, J., Y. Yan, T. P. J. Garrett, J. Liu, D. W. Rodgers, R. L. Garlick, G. E. Tarr, Y. Husain, E. L. Reinherz, and S. C. Harrison 1990. Atomic structure of a fragment of human CD4 containing two immunoglobulin-like domains. *Nature* **348**:411–18.

Waring, M. J. 1981. DNA modifiers and cancer. *Ann. Rev. Biochem.* **50**:159–92.

Watson, J. D., and F. H. C. Crick 1953. Molecular structure of nucleic acids. A structure for deoxyribonucleic acid. *Nature* **171**:732–38.

Weinberg, R. A. 1985. The action of oncogenes in the cytoplasm and nucleus. *Science* **230**:770–76.

Weis, W., J. H. Brown, S. Cusack, J. C. Paulson, J. J. Skehel, and D. C. Wiley 1988. Structure of the influenza virus haemaglutinin complexed with its receptor sialic acid. *Nature* **333**:426–31.

Weiss, M. S., T. Wacker, J. Weckessen, W. Welte, and G. E. Schulz 1990. The three-dimensional structure of porin from *Rhodobacter capsulatus* at 3Å resolution. *FEBS Letts.* **267**:268–72.

White, P., K. J. Buchkovich, J. M. Horowitz, S. H. Friend, M. Raybuck, R. A. Weinberg, and E. Harlow 1988. Association between an oncogene and an anti-oncogene: The adenovirus E1A proteins bind to the retinoblastoma gene. *Nature* **334**:124–29.

Wigley, D. B., G. J. Davies, E. J. Dodson, A. Maxwell, and G. Dodson 1991. Crystal structure of an N-terminal fragment of the DNA gyrase B protein. *Nature* **351**:624–29.

Wiley, D. C., and J. J. Skehel 1987. The structure and function of the hemaglutinin membrane glycoprotein of the influenza virus. *Ann. Rev. Biochem.* **56**:365–94.

William, L. D., M. Egli, Q. Gao, P. Bash, G. A. Van der March, J. H. Van Boom, A. Rich, and C. A. Frederick 1990. Structure of nogalmycin bound to a DNA hexamer. *Proc. Nat. Acad. Sci. USA* **87**:2225–29.

Wilson, I. A., J. J. Skehel, and D. C. Wiley 1981. Structure of haemaglutinin membrane glycoprotein of influenza virus at 3Å resolution. *Nature* **281**:366–73.

Wlodawer, A., M. Miller, M. Jaskolski, B. K. Sathyanarayana, E. Baldwin, I. T. Weber, L. M. Selk, L. Clawson, J. Schneider, and S. B. H. Kent 1989. Conserved folding in retroviral proteases: Crystal structure of a synthetic HIV-1 protease. *Science* **245**:616–26.

Wüthrich, K. 1986. *NMR of Proteins and Nucleic Acids.* John Wiley and Sons, New York.

Wüthrich, K. 1989. The development of nuclear magnetic resonance spectroscopy as a technique for protein structure determination. *Acc. Chem. Res.* **22**:36–44.

Yang, W., W. A. Hendrickson, R. J. Crouch, and Y. Satow 1990. Structure of ribonuclease H phased at 2Å resolution by MAD analysis of the selenomethionyl protein. *Science* **249**:1398–1405.

Zhang, R.-g., A. Joachimjak, C. L. Lawson, R. W. Schevitz, Z. Otwinowski, and P. B. Sigler 1987. The crystal structure of the trp repressor at 1.8Å shows how binding of tryptophan enhances DNA affinity. *Nature* **327**:591–97.

Zhang, J., L. S. Cousens, P. J. Barr, and S. R. Sprang 1991. Three-dimensional structure of human basic fibroblast growth factor, a structural homolog of interleukin 1β. *Proc. Nat. Acad. Sci. USA* **88**:3446–50.

Zhu, X., H. Komiya, A. Chirino, S. Faham, G. M. Fox, T. Arakawa, B. T. Hsu, and D. C. Rees 1991. Three-dimensional structures of acidic and basic fibroblast growth factors. *Science* **251**:90–93.

Index